Bioethics and the Environment

Bioethics and the Environment

A Brief Review of the Ethical Aspects of the Precautionary Principle and Genetic Modified Crops

Luis G. Jiménez-Arias

www.librosenred.com

C.E.O.: Marcelo Perazolo
Managing Editor: Ivana Basset
Cover Design: Daniela Ferrán
Interior Design: Javier Furlani

Design, typesetting, and other prepress work by LibrosEnRed
www.librosenred.com

First English Edition - Print on Demand

ISBN: 978-1-59754-380-4

LibrosEnRed©
A trade mark of Amertown International S.A.
editorial@librosenred.com

TABLE OF CONTENTS

To my mother, Bettina, from whom I learned to love creation

Acknowledgments

Bioethics and the Environment is the result of my work during my studies at *Regina Apostolorum* University, in Rome, Italy. It needed the help of many professionals in this field. I do thanks specially to Dr. Carlos Petrini, from the Bioethical Unit at the Centro Nazionale di Epidemiologia, Sorveglianza e Promozione della Salute, Istituto Superiore di Sanità, Rome. To Shirley Schley for her kindly assistance and to all the faculty members from *Regina Apostolorum* school of bioethics. To all those professionals who kindly provided me with theirs works and citations specially Dr. Karen-Beth G. Scholthof, Ph.D. Associate Professor, from the Department of Plant Pathology and Microbiology Texas A&M University. To Dr. Philippe Grandjean, MD, PhD, from the Occupational Health Program, Harvard School of Public Health.

PREFACE

There are two things that I love more: philosophy and nature. The first help us to understand the second and both help us to know God. Since the publication of *Silent Spring* in 1962 by Rachael Carson to today, thousands of papers and books on environmental ethics have been written. The debate is polarized on what is convenient and should be done and what a man of conscience should avoid, especially in regard to the impact of agriculture on the environment, whose ultimate goal is to provide a better standard of life to the world population. Here's the dilemma: while traditional agricultural systems appear to be unsustainable, due to the environmental impact, the transgenic production seems an oasis in the desert that could help us to make more sustainable the agricultural production. To reconcile the ethical principles governing the scientific activity with human development is one of the most frequent ethical concerns in dealing with protecting the environment. This work is just an attempt to clarify some of the precepts —such as the Precautionary Principle— hat seems to govern many field of scientific research especially transgenic crops and ultimately human development.

INTRODUCTION

Now as ever before, we have to face ethical challenges in light of scientific and technological developments. The genomic intervention of man in plants, animals and microorganisms has some risk for them and, consequently, for humankind. On the other hand, if people were to not interfere in their genome, it might have some unfavorable consequences for human survival as well as detriment for the environment. Risk estimation of human intervention in other living organisms should be fairly evaluated against the risk of not intervening. As we develop this issue further, one essential question arises: What offer more risk: intervening or not intervening?

But before answering this question, we have to know some facts related to a wider scope of this topic. The first fact is that there are already millions of people starving or suffering from nutrition deficiencies around the world. The second fact is that traditional intensive agricultural systems based on pesticides and agrochemicals seems to have failed because they might not be sustainable and because most existing pollutants come from agricultural production for feeding humans and animals. The third fact is that most technological advances that can help solve hunger around the world come from research done in developed countries and might not be adopted in developing countries. The last fact is that these technologies do not necessarily transfer, at least at the necessary speed, to the poorest countries. It is here that the cooperation principle and the so-called precautionary principle (PP), which seems to be gaining recognition, come together. It is here where bioethics and bio-law have the opportunity to find answers to many questions that require an interdisciplinary dialogue.

Safe food, clean water and unpolluted air are basic human needs but, if they are polluted to some degree, they can also become hazardous to human and animal health.

The paradox is that the deficiency of precautionary guides and actions might result in irreversible or serious damage to ecosystems and human health, or the contrary, the wrong application of precautionary measures, based on suspicious risk, might also result in these same problems.

As new ideas, concepts and principles are established to conceptualize new events; economic, political and ideological groups come out to debate them. One of the most environmental concepts that are in the spotlight is the so called Precautionary Principle (PP). The questions surge: is PP an ethical principle? Is it a practical idea? Is PP being overused, misused or misinterpreted? Are we actually destroying our environment? Can we apply the Principle of Cooperation, so widely applied in medical ethics, to a situation which concerns the essential needs of humans, such as food and water?

In this book, we will explore some areas of knowledge and human activity such as agriculture, human health, the effect of people on the environment and the subsequent effect of the environment in return on humans.

In order to explore these areas of knowledge, this book is divided into four chapters. The first chapter is about the origin of the Precautionary Principle (PP), the political and social need for a new concept to stop environmental damage. The second chapter is about the search for a solution to humankind's nutrition problem. The third chapter is about the environment and the ecosystem, biodiversity, air pollution and the ethical debate concerning humans and their relation with the environment. Finally, the last chapter, of this book, in about the most relevant issues that involve the PP.

Chapter I
I-The Concept of the PP

I.1 Development of the PP concept:

The modern concept of the PP seems to come from the German word *"Vorsorgeprinzip,"* introduced in the 1930's and then extended in the 70's as environmental policy. [1] Later, in 1984, the term "Precautionary Principle" was included in the Convention on the Protection of the Nordic Sea and followed by many other agreements like the Rio Declaration and for the European Union "Precaution" is part of their primary and secondary European legislation. [2] [3]

Rio Declaration:

In 1992, the Rio Declaration came about when the United Nations Conference on Environment and Development met at Rio de Janeiro, Brazil and reaffirmed the Decla-

---

[1]   BOECHMNER C.S. 1994. *The Precautionary Principle in Germany.* In *Interpreting the Precautionary Principle.* T. O'RIORDAN and J.CAMERON (Eds). London. P.31-60.

[2]   VAN DER HAEGEN T. 2003. *A view of the precautionary principle in food safety.* Speech at American Branch of the International Law Association. New York, October 23-25, 2003. http://www.eurunion.org/News/speeches/2003/031023tvdh.htm

[3]   EUROPEAN COMMUNITY. 2002. General Food Law - Precautionary Principle Regulation. EC178. http://ec.europa.eu/food/food/foodlaw/precautionary/index_en.htm

ration of the United Nations Conference on the Human Environment, adopted at Stockholm on June, 1972[4]. This declaration rests on 27 principles. The Rio Declaration aims to set up a new and fair global alliance through the encouragement of cooperation among countries as well as all levels of society in order to set the basis for international agreements. Rio also understands clearly that the environment is an international issue, that what is done in one part of the planet affects the other, that is why this declaration aims to protect the global environment while taking into consideration the interest of all parties involved in the process.

At Rio, the precautionary approach (PA) was used as synonym for the PP, where world participants adopted Principle 15 that says: "In order to protect the environment, the Precautionary Approach shall be widely applied by states according to their capabilities. Where there are threats of serious or irreversible damage, lack of full scientific certainty shall not be used as a reason for postponing cost-effective measures to prevent environmental degradation".[5]

This principle is clear in the term environment; it aims to protect not exclusively the environment since Principle 1 mentions that "Human beings are at the centre of concerns"[6]. This statement has three other characteristics, the first one is relative to the countries, it encourages countries to do so, the second one the "threat" of damage and the last one is that Rio introduced the cost-effective measures.

---

4   UNITED NATIONS. 1972. Conference on human environment. 1972. Report. Stockholm, June 5-16.

5   UNITED NATIONS. 1992. Conference on environment and development (UNCED). June 3-14. Rio de Janeiro, Brazil. Principle 15.

6   UNITED NATIONS. 1992. Conference on environment and development (UNCED). June 3-14... Rio de Janeiro, Brazil. Principle 1.

Wingspread Conference Definition

The Wingspread declaration definition of the PP is one of the most mentioned, this conference, held in 1998 in the United States, was a multidisciplinary gathering of scientists, philosophers, lawyers, environmental activists and others caring for the human health and the environment.

The term precaution, in the Wingspread declaration, seems to be the reaction to threats of harm, according to Wingspread, cautionary actions might be more important than scientific findings. The literal definitions for PP in Wingspread declaration is "When an activity raises threats of harm to human health or the environment, precautionary measures should be taken even if some cause and effect relationships are not fully established scientifically." [7]

The Wingspread declaration seems to be more explicit than the United Nations Conference on Environment and Development (UNCED) [8] and goes beyond it. This declaration includes the term "human health" and makes a clear separation between human health and the environment. The Wingspread declaration mentions the word "activity," which is not specific about a human or non human activity; this activity is a cause and might have an effect, according to Wingspread declaration even if the effect can not be established precautionary measures should be taken. The Rio definition seems to link human health to the environment and focus its attention in this relation human-environment as "Human beings are at the centre of concerns for sustainable development. They are entitled to a healthy and productive life in harmony with nature." [9] Furthermore, in Wingspread Conference precautionary measures are to be taken in cause-effect relationships

[7]   WINGSPREAD CONFERENCE. 1998. Racine, Wisconsin.
[8]   UNITED NATIONS. 1992. Conference on environment and development (UNCED). June 3-14. Rio de Janeiro, Brazil. Art. 27.
[9]   UNITED NATIONS. 1992. Conference on environment and development (UNCED). June, 3-14. Rio de Janeiro, Brazil. Principle 1.

different from human-environment. This cause-effect relation is not only to prevent harm to the environment; it is also applicable in terms of harm to human health.

CARTAGENA PROTOCOL

Following Wingspread Conference, on January 2000 and taking into consideration Rio principle 15, Cartagena Protocol aims to establish different levels of protections, in other words tries to make Principle 15 practical. Cartagena declares: "In accordance with the precautionary approach contained in Principle 15 of the Rio Declaration on Environment and Development the objective of this Protocol is to contribute to ensuring an adequate level of protection in the field of the safe transfer, handling and use of living modified organisms resulting from modern biotechnology that may have adverse effects on the conservation and sustainable use of biological diversity, taking also into account risks to human health, and specifically focusing on transboundary movements."[10]

This level of protection refers mainly to the transfer of living modified organisms (LMOs) in order to protect the biodiversity. In order to implement the levels of protection, Cartagena proponed the regulation of movement of LMOs between the participant parts. Cartagena Protocol takes into account human health however this is not the focus of this Protocol. The definition established by the Cartagena Protocol goes beyond the previous ones by extending the potential harm to animal or plant health.

THE CANADIAN PERSPECTIVE

According to the Canadian Government, the precautionary approach/principle is a distinct approach within science-based

---

[10]   CARTAGENA PROTOCOL ON BIOSAFETY. 2000. Cartagena, Colombia. Art.1. http://www.biodiv.org/biosafety/ratification.asp

risk management. It seems to recognizes that the "absence of full scientific certainty shall not be used as a reason to postpone decisions where there is a threat of serious or irreversible harm"[11]. Canada adopts the term "Precautionary Approach" to describe the concept of precaution, the term "precautionary principle" seems to lose ground in the Canadian perspective. In legal terms the use of Precautionary Approach instead of precautionary principle can have some influence in local law as well as in international law. It looks like Canada does not fully accept the word "principle" for the PP[12]. In practice, the Canadian Government has established some guides for the application of the PP. Terminology for the same concept is very important because it might change the legal status and the relevancy to deal with a specific issue. For instance, "Respect for Nature", that is also being used, seems to call for awareness or is a calling to consciousness; also "Respect for nature might imply an attitude of prudence. This Canadian perspective introduces something new in the development of the PP, according with this perspective the levels of protection are relative to the scientific data "The scientific evidence required should be established relative to the chosen level of protection".[13]

PP in the European Union

The European environmental law, according to the Treaty on European Union (1992), takes the PP as the base. The Europe-

[11]   GOVERNMENT OF CANADA. 2001. A Canadian Perspective on the Precautionary Approach/Principle Discussion Document, September.

[12]   LEE S., BARRETT K. 2002. Comments on: A Canadian Perspective on the Precautionary Approach/Principle Discussion Document. http://www.sehn.org/canpre.html

[13]   GOVERNMENT OF CANADA. 2001. A Canadian Perspective on the Precautionary Approach/Principle Discussion Document, September.

an Commission states that "where scientific evidence is insufficient, inconclusive or uncertain.""[14] the PP should be applied. In this sense the PP is applied when there is not enough evidence or the evidence is open to doubt or doubtful; furthermore, this treaty mentions the "potential dangerous effect" no only on the environment but also on humans, animals and plant health. Here we can observe a relationship between the "chosen level of protection" and the "potential dangerous effects". If the chosen level of protection might not cover, in a reasonable way, the health of living organisms then the PP should be apply.

On February, 2000 a special commission of the European Union, following the Cartagena Protocol on Bio-safety, adopted the Communication on the Use of the Precautionary Principle. The Communication highlights the fact that the PP should be considered within a structured approach to risk analysis, and it is particularly relevant to risk management.[15] Basically, the Communication states that the PP "forms part of a structured approach to the risk analysis, as well as being relevant to risk management." This risk analysis is specifically related to the lack of scientific evidence. These evidences are insufficient, inconclusive or uncertain or if the preliminary scientific evaluation indicates that there are reasonable grounds for concern of potentially dangerous effects on the environment, human, animal or plant health.

PP in the World Health Organization

The PP seems to lack clear and universally accepted definitions and actions. It is not universally clear how countries

---

[14]   EUROPEAN COMISSION. 2000. Communication on the precautionary principle. http://ec.europa.eu/dgs/health_consumer/library/pub/pub07_en.pdf.
[15]   EUROPEAN COMMUNITY. 2000. Communication from the commission on the precautionary principle. http://ec.europa.eu/environment/docum/20001_en.htm

should apply the PP. Most recently, a World Health Organization (WHO) international workshop was held to develop a common framework for application of the PP to possible health[16] issues. PP as defined by WHO is not a principle but a concept: "a concept that allows flexible approaches to identifying and managing possible adverse consequences to human health, even when it has not been established that an activity or exposure constitutes harm to health."[17] In this context, the problem seems to be a misunderstanding between precaution, as a principle, and preventive measures.[18]

In order to understand what is precautionary measures and what is preventive measures one must distinguish between both concepts and definitions. According to Cameron "Preventive standards may be precautionary or non-precautionary in certain degrees, but precautionary standards, while able to vary the degree of prevention, cannot be non-preventative."[19] The reason is found in the relation between precaution and certainty, precaution is applied when there is a lack of certainty. In other words when the cause/effect relationship is unknown and there might be a risk of harm to the environment. On the contrary if there is evidence or certainty on the environmental risks then the measure is preventive. Therefore the difference between certainty and uncertainty is between knowing the cause/effect and not

---

[16]   WORLD HEALTH ORGANIZATION. 1946. Constitution. The definition of health is a "state of complete physical, mental and social well being and not merely the absence of disease or infirmity." http://en.wikipedia.org/wiki/Health
[17]   EUROPEAN COMMUNITY AND UNITED STATES NATIONAL INSTITUE FOR ENVIROMENTAL HEALTH SCIENCES. 2003. Application of the Precautionary Principle. Luxembourg.
[18]   CAMERON J. 1995. The Status of Precautionary Principle in International Law. In T. O'Riordan and J. Cameron (eds.) Interpreting the precautionary principle. Cameron May Ltd., UK, 315 p.
[19]   Idem.

knowing it. Acting can be preventive or precautionary. In the face of uncertainty, however, the precautionary principle, like the Vorsorgeprinzip, allows for the state to act in effort to mitigate the risks. Put best, "the precautionary principle stipulates that where the environmental risks being run by regulatory inaction are in some way uncertain but non-negligible, regulatory inaction is unjustified".[20]

PP in the Social Doctrine of the Catholic Church

Precautionary Approach is recommended in the guidelines for food safety however, Precautionary Approaches should be focus on uncontrolled risk and not in "working techniques coming from the new technologies"[21]. The Holy See base his support to precautionary measures on the "safety and quality of the agro-food crops, as an expression of a balanced relationship between the order of the Creation and human's activities" [22] But also there is some concern on the coordination among countries to put into effect the regulations. The safety and quality involves the prevention in the production phase or risk assessment and the risk management. [23]

The Principle of Responsibility, applied to the environment seems to find an eco in the teaching of the Catholic Church, furthermore, there is a recommendation that this principle should be put into legislation "…find adequate expression on a juridical level"[24] and apply this to local and international

---

[20]   Idem.

[21]   JOHN PAUL II. 2000. Speech on the occasion of the Jubilee of Agricultural World. November 11.

[22]   Idem.

[23]   VOLANTE R. 2003. Intervention by the Holy See at the 32nd session of the conference of FAO. Rome, December 3.

[24]   PONTIFICAL COUNCIL FOR JUSTICE AND PEACE. 2005. Compendium of the social doctrine of the church. 10:2, 468. Libreria Editrice Vaticana, Vatican.

law. International community drew up uniform rules that will allow States to exercise more effective control over the various activities that have negative effects on the environment and to protect ecosystems by preventing the risk of accidents.[25] This position of the Catholic Church aims at global legislation to protect the environment as well as human health and is based in a previous statement from Pope John Paul II[26] "The State should also actively endeavor within its own territory to prevent destruction of the atmosphere and biosphere, by carefully monitoring, among other things, the impact of new technological or scientific advances [and] ensuring that its citizens are not exposed to dangerous pollutants or toxic wastes".[27]

The PP as understood by the Holy See is not an orthodox principle of applying rules instead it calls for "certain guidelines aimed at managing the situation of uncertainty". This is the case when the scientific data is "contradictory or quantitatively scarce" and authorities have to make decisions, then it may be appropriate to base evaluations on the PP. Also these guides to manage situations of uncertainty put into effect in the lack of valid scientific data should be temporary and keep the Principle of Proportionality in reference to other decisions that are made for other risks.

The Holy See calls for the virtue of prudence when dealing with the PP. This means that decisions of intervention or not "based on the precautionary principle require that decisions be based on a comparison of the risks and benefits foreseen for the various possible alternatives."[28] Uncertainty will be always

---

[25]   Idem.

[26]   JOHN PAUL II. 1990. Message for the 1990 World Day of Peace. 9: AAS 82, 152.

[27]   Idem

[28]   PONTIFICAL COUNCIL FOR JUSTICE AND PEACE. 2005. Compendium of the social doctrine of the church. 469. Libreria Editrice Vaticana, Vatican City.

present since it is almost impossible to know all risk associated to any specific activity.

This position of the Holy See on the PP is a lot more soft and practical that many others interpretations of the PP. The differences between the Holy See proposal and other states in summary are: a call for the responsibility, a call for prudence, a fair balance between benefits and risk, the to use of PP in the benefit of both human health and the environment, this include the flexibility of the principle in applying guidelines as new data is obtained and finally it calls for transparency in the process.

OTHER DEFINITIONS

Some authors call to treat the PP with caution, there are two main reasons: First, they doubt that the PP is a principle, and, secondly, because it has not been explicitly defined.[29] Since the PP is mainly mentioned in environmental protection and genetic manipulation discussions, S. Holm and J. Harris [30] believe that more proponents of the PP will accept the following definition: "When an activity raises threats of serious or irreversible harm to human health or the environment, precautionary measures that prevent the possibility of harm (for example, moratorium, prohibition) shall be taken even if the causal link between the activity and the possible harm has not been proven or the causal link is weak and the harm is unlikely to occur."

The problem with this definition is that it might be balancing evidence in a specific way. This means that the PP, as stated, might result in a logical problem. It is say that scientific knowledge is based in facts, therefore evidence is necessary. In

---

[29]   HOLM S., HARRIS J. 1999. Precautionary principle stifles discovery. Nature, 400: 398.
[30]   dem.

the specific case of PP evidence is an obligation for one side. Evidence is what provides reason for believing something. Statistical analysis is a scientific tool that does not deal with the truth, but with facts. The truth belongs to philosophy, not to science. Weight given to evidence is ordinarily thought to be a "function of its epistemic warrant (the degree to which we have reason to believe the evidence), even in cases where the evidence has the same epistemic warrant".[31] Then according to Holm and Harris, PP cannot be a valid ethical principle. Furthermore, they do not consider it a valid tool to evaluate evidence because if PP is applied as stated then it might distort our beliefs about the world and could lead us to hold false beliefs accepting something as truth when it is false.

I.2. TESTING PP AS A VALUABLE PRINCIPLE

Is PP an ethical principle? In order to answer this question, one must know that a principle, in short, is a departure point. Ethical principles as well as moral theories depend on the human concept that inspires them; principles and moral theories can not be evaluated without this. Therefore, the human vision (or the lack of it) of humankind's final end (great or small), exercises enormous influence on the ethical/moral theory. It is said that the must practical question you can ask anyone is: What is your philosophy of life? No one can just take a concept and convert it into a principle. According to W.B. Smith, "principles are articles of faith and virtues are rational principles".[32] Therefore, principles can evaluate the good or the wrong of human acts and human acts proceed from the free will of humans with the knowledge of the end goal.

---

[31]   Idem.
[32]   SMITH W. 2001. Class notes, Saint Joseph's Seminary. Dunwoody, New York.

Principles must be practical and should lead to the right judgment. They are products of reason, and, even for utilitarian philosophers, an ethical judgment must be practical. This means that if an ethical principle is unpractical, it cannot be principle of all. P. Singer, in the chapter: About Ethics, in his book *Practical Ethics* wrote: "an ethical judgment that is not good in practice must suffer from a theoretical defect as well, for the whole point of ethical judgment is to guide practice."[33] And Jordan and O'Riordan mentioned: "...precaution lacks a specific definition, and, as yet, it cannot prescribe specific actions or solve the kind of moral, ethical and economic dilemmas which are part and parcel of the modern environmental condition."[34]

From the above definition of principles and practical ethics, we can find some clues to evaluate PP as a principle or as a mere concept or set of concepts not yet well defined. The so-called environmental movement, which claims to be ecologist, has proceeded by capturing ideas and transforming them into principles, guidelines and points of leverage. For instance, agricultural sustainability is one such idea now being reinterpreted and implemented in the aftermath of the Rio Conference, so is the PP. PP is suffering from all kinds of manipulations to make a mere principle out of it. This means taking social norms and converting them into a universal, well-accepted "principle". The problem is that principles are what they are —principles, and not abstract concepts, or agreements. Furthermore, principles are not mere norms but ethical norms.

---

[33]   SINGER P. 1993. Practical Ethics. Cambridge University Press, Cambridge. p2.

[34]   JORDAN A., O'RIORDAN T. 1998 The Precautionary Principle in Contemporary Environmental Policy and Politics. Paper prepared for the Wingspread Conference on 'Implementing the Precautionary Principle', Racine, Wisconsin.

PP AND THE ENVIRONMENTALISTS

In terms of the environmental polices and the influence of the environmentalists the PP has its consequences, to start with if we accept the that PP is a principle it created the baseline to launch of the ideological platform for the environmentalist, it means of the eco-centric culture, it gives a philosophical and ethical tool to the green extremist and they might feel protected and use the PP as shield for their ideological attack. Furthermore, just by accepting that the PP is a principle it might challenge the economical, social and political system.

The moral objective of the environmentalist movements does not appear to be bad, since to protecting the environment is a good human act, even more it is a moral obligation. The problems are the means used by some environmentalist to reach their objectives. It is a matter of common sense to create a balance between human development and environmental protection; sustainable human and economical development can be reached only if environmental and natural resources are protected. To use the PP as a tool to stop economical and human development is therefore wrong. Environmental protection can not be approached as an article of faith, can not become a religion or a matter of belief it is in fact a matter of education, of global cultural conscience, that should be incorporated into global and local law.

If nature has an intrinsic moral value then, it could have intrinsic rights, as some extreme environmentalist's aims to give it. Giving nature intrinsic moral values and rights result in a serious logical problem from both biological and ethical perspective. Since bioethics departs from biological facts to give nature an intrinsic moral value just because it is nature, is to go against the rules of biology, especially those widely accepted in science, about the origin, evolution and adaptability and competitiveness of the species.

PP, as stated at Rio, and as interpreted by some scholars, has serious limitations on the development of science and pubic discussion. The efficacy of this principle seems to have more to do with political and environmental movement than with logical ethical debate. Furthermore, it has very little to do with ethics because, as mentioned by Riordan and O'Brian, "Nonetheless, the precautionary principle has much efficacy because it captures an underlying misgiving over the growing technicalities of environmental management at the expense of ethics and open dialogue..."[35]

Side effects on adopting the PP

According to Holm and Harris[36] if any given principle obstructs the development of science, based in beliefs and not in reason or theoretical possibility of harm, then it cannot be a valid rule for rational decisions, they mentioned that PP is not a principle of "rational choice." For them the greatest uncertainty about genetically modified (GM) plants and the possible harmfulness existed before anybody had yet produced one. Also these authors argue that if we follow PP, we can not proceed any further, and no data will be available to demonstrate if an activity is harmful or not to human health or the environment, therefore, according to these authors PP in an irrational principle.

Hans Labohm concludes that PP "can lead to technological stagnation, barriers to trade and loss of human lives"[37]. In his work, Labohm goes over the consequences of over-using the

---

[35] JORDAN A., O'RIORDAN T. 1998. The Precautionary Principle in Contemporary Environmental Policy and Politics. Paper prepared for the Wingspread Conference on 'Implementing the Precautionary Principle. Racine, Wisconsin, January 23-25.

[36] SOREN H., HARRIS J. 1999. Precautionary principle stifles discovery Nature, 400: 398.

[37] LABOHM H. et al. 2003.Cannons and Canons: Clingendael Views of Global and Regional Politics. Van Gorcum Ltd.415p

better safe than sorry approach. He deeply explores PP, and also concludes that proponents of PP do not take into account the benefits of new products and processes. He does agree that nothing is risk-free. He concludes that economic development depends upon the innovation of technology that is penalized by PP, while PP does accept the risks associated with existing activities and products. The author based his work on facts about the over-precaution of PP, such as the emergence of malaria in some developing countries because of the campaign against the use of the pesticide DDT to combat malaria[38]. Another fact is that PP has aroused suspicions of protectionist motives and trade tensions with the U.S. because of the moratorium on GM food of the European Community (EC).

The attitude of the EC toward GM food is affecting Africa, where millions are dying of starvation. According to Kershen, Africa's position on GM crops is more likely based on fear since Africa depends on EC aids and the export of agricultural products to the EC[39].

The belief that humans have a nefarious influence on climate dynamic seems to be becoming an "article of faith."[40] Gerhard and Yannacone[41] mentioned some facts about how nature itself influences weather change; these facts are driving out the belief that it is human activity that is responsible for environmental changes. Also according to these authors; politicians, journalists and scientists are more cautious when dealing with affirmative statements related to the influence of human activity on climate changes.

---

[38]   LEWIS K. 2006. Condemned to die' by malaria, Ugandans plead for DDT use. http://www.fightingmalaria.org

[39]   KERSHEN D L. 2002. Law and Horror. http://www.tcsdaily.com/Article.aspx?id=122002M

[40]   GERHARD L. C., YANNACONE V. J. Jr. 2004. Invoking a Real Precautionary Principle. .- http://www.techcentralstation.com

[41]   Idem.

New data seem to reveal that the influence of human activity on climate changes is not so. It looks like some environmental movements still support the "articles of faith of anthropocentric climate change"[42] and they have found in PP a shelter to hide from reality. Humans are responsible in some extent for weather changes, but the grade of influence of human activities in these changes is still not clear. As mentioned by Gerhard and Yannacone[43] "Their Precautionary Principle requires current action to mitigate speculative future impacts regardless of the present consequences —intended or unintended."

Setting the cost-benefit of the PP

For Raffensperger[44] PP invites us to return to an ethical focus on science. He agrees with the idea that PP is a way to regulate science. Furthermore referring to regulation of professionals in fields like law or medicine, this author argued: "What if scientists shared that same obligation to use their skills for the good, *pro bono*?

But, is it therefore morally right to let people die or to cause more environmental damage just because we are not 100 percent sure that new technologies are 100 percent safe? The right answer might be an absolute not. Statistically, in every human action, there is a degree of risk. Nature itself is risky; for instance, the risk associated with drinking contaminated water, the risk of eating any type of contaminated food, the risk of breathing polluted air. Paradoxically, precaution follows the steps of sustainability; it is being integrated into

---

[42]   Idem.

[43]   Idem.

[44]   RAFFENSPERGER C., TICKNER J. (Eds.). 1999. Protecting Public Health and the Environment: Implementing the Precautionary Principle. Island Press. Washington DC.

modern environmentalism. In excess, precaution is as risky as insufficient precaution. If technology and scientific research was done only when there is no risk, then it would never be done, because there is never a hundred percent certainty about the consequences of the activity.

Doing nothing is not, in itself, without risk. Doing nothing does not go, by any means, in favor of the poor. This omission is frequently found in the proponents of PP because the possible benefits of research are not taken into account.[45]For instance, based on the Widespread Conference,[46] activists in the U.S. are calling for a complete ban on GM food. Their principle argument is that science cannot prevent all the consequences from the introduction of GM food to the environment.

Sustainability vs. PP.

For some scholars neither sustainability nor the PP are ethical principles. Instead, in 1992 Norton[47] calls sustainability "a set of principles, derivable from a core idea of sustainability, but efficiently specific to provide significant guidance in day to day decisions and in policy choices affecting the environment." Therefore they are not ethical or moral principles, furthermore they do not seem to be well defined concepts.

According to Tickner[48], rather than a well-defined or stable principle, sustainability seems to be a group of concepts wrapped into one word. This same conceptualization might

[45]   CAROLINA C. *et al.* 2004. Environmental biosafety and transgenic potato in a centre of diversity for this crop. Nature, 432, 222 – 225.
[46]   WINGSPREAD CONFERENCE. 1998. Racine, Wisconsin, USA.
[47]   NORTON B. 1992. Sustainability, Human Welfare and Ecosystem Health. Environmental Values 1: 97-112.
[48]   TICKNER J.A. 2004. Children in Their Environments: Vulnerable, Valuable & at Risk: The Need for Action. EEA/WHO/Collegium Ramazzini A one-day Science/Policy Workshop, Budapest. June.

be happening to the PP. The PP reflects the early thoughts of the German *Vorsorgeprinzip*, who intended it to be a social planning principle to achieve sustainability and promote safer technologies.[49]

But for Bodley point of view, sustainability, as well as the PP, are "dangerously successful" in the sense that both are vague concepts. Sustainability seems to affirm that our modern industrial civilization (including economical, social, political and agricultural systems) is not sustainable on the other hand PP seems to evoke us that precautionary measures should be taken because, so far, none (precautionary measures) are being taken.

In sustainable agriculture, for instance, one of the goals is to reduce the input of chemicals into farms, while organic agriculture, which claims to be the most sustainable system, aims to completely avoid synthetic chemicals. It is almost impossible to grow most extensive crops, including GM crops, without the help of agricultural chemicals. Most genetic engineered (GE) crops would help make the system more sustainable because the aim is to reduce their use. However, GE crops are not considered natural, and are even considered by some to be dangerous to the environment and consumers.

But is sustainability a new concept? According to Bodley[50], indigenous people, normally perceived as primitive by anthropologists, have evolved sustainable use of natural resources for hundreds of years. He claims that some tribes and civilizations have realized the value of conservation and that preserving nature is part of these so-called "primitive" societies. Furthermore, he maintains that perhaps they preserve nature because they already know the consequences of destruction. Many anthropologists believe that the Maya Civilization in north-

---

[49] Idem.
[50] BODLEY J. 1990. Victims of Progress. Mayfield, Mountain View, California. 261 pp.

ern Central America was extinguished because of the overuse of agricultural land. As a consequence of natural disasters and based on their own cultural experience indigenous people in Central America have a great religious devotion to nature, in the past it included human sacrifices. Bodley wanders if this religious devotion to nature is ignorance, or is it based on their ancestors' experiences?

Many of the so-called "primitive societies" might not be so primitive. In Costa Rica some ethnic groups have been practicing both sustainability and the PP; as a consequence they are the major exporters of organic cocoa, bananas and many other exotic fruits. We do not have to teach them how to take appropriate care of their environment, by their acts they teach us (without knowing it) what is precaution and what are sustainable agricultural systems. No wonder Bodley mentioned that "our modern civilization, which often believes sustainability is a new concept, should understand that sustainability is just a new word which is a product of the lack of humility because we do not want to accept that the so-called primitive civilization are not so primitive".[51]

I.2. PRECAUTION AND PREVENTION

PRECAUTION

In the field of health, for Gradjean,[52] precaution is common sense and should be the basis of any healthy decision. He maintains that "precaution is not merely a methodological system; rather, it is directly related to any decision in

---

[51]   Idem

[52]   GRANDJEAN Ph. 2004. Implications of the Precautionary Principle for Primary Prevention and Research. Annual Review of Public Health, 25:199-223.

public or individual health." For this author it is therefore acceptable that precaution is related to the virtue of prudence, and prudence would lead us to the conclusion that the benefits of any doubt should fall upon the patient or *in dubio pro salus*.

Precaution seems to have low and high limits. Too little or too much precaution might be seen as risky. However, not all authors share this position; some believe it is not dangerous, even in excess. *"Abundans cautela non nocet"*, was affirmed by Baker[53] in reference to the possible toxicity of lead in the eighteenth century to sweeten the taste of sour wine. Can abundant precaution cause no harm? In order to answer this question we can take the case of AIDS research, in the case of vaccine development to much precaution could lead to the extermination of people in some African areas causing a lot of harm, in this case cost/benefit approach should be applied. Also for instance too much precaution in researching and releasing some Genetically Modified Organisms (GMO's) can speed up malnourishment and furthermore contribute to starvation or to an increase in blindness due to a lack of vitamin A, this is the case with Golden Rice, that at a normal intake provides a large part of the total vitamin A that is needed to prevent blindness.[54]

PREVENTION

There seems to be confusion between prevention and precaution. They might have the same meaning in some languages, but, at

---

[53]   BAKER G. 1772. An Inquiry concerning the cause of the endemically colic of Devonshire. Med. Trans. R. Coll. Phys, 1;175-256.

[54]   NOLUTSHUNGU T. 2006. Question of Life or Death in Africa. The Standard (Hong Kong), Feb. 8. http://www.goldenrice.org/Content4-Info/info7_actuality.html

least in English, which is the official languages of most documents, there is a difference in meaning between the two words.

It is very important to distinguish between potential risk (A factor, thing, element, or course involving uncertain danger; a hazard) and proven risks related to the parallel distinction between precaution (Preparation and disposition that is done in advance to avoid a risk or to execute something. An action taken in advance to protect against possible danger, failure, or injury; a safeguard). For instance, the prevention of nuclear war vs. the precaution concerning GMO. Potential risks are related to precaution in the same manner that prevention is related to proven risks.

Frequently, precaution and prevention are used indistinctly, creating confusion between both words. The reason for this confusion is that potential risk is assumed to be a proven risk. If there is a proven risk, such as contracting malaria, then preventive measures should be taken, for instance, the use of mosquito repellents, pills and other measures. However, if we go to the tropics and malaria is not found in all areas but, we are afraid of getting malaria, such measures, simply because of fear, should not be taken.

In other words, one often thinks that the potential risks, which are not very probable, are proven risks, and, in this case, precaution is assimilated as prevention. Statistically speaking, the probabilities do not have the same nature. In the case of the precaution, it is a matter of the probability that the hypothesis is exact; in the case of the prevention, danger is established, and it is a matter of the probability of the accident occurring.

In addition, potential risks, in spite of their hypothetical character, can have a high realization probability. In practice, nevertheless, the precaution can be understood as the extension of the methods applied to prevent the uncertain risks.

I.3 PP AND RISK MANAGEMENT

LEVELS OF PROTECTION

The PP referrers to circumstances in which there is not enough scientific evidence or there is uncertainty about the safety of human actions. In other words, "where there are still some concerns and insufficient scientific data to evaluate potentially dangerous effects on the environment, human, animal or plant health".[55]

According to Gradjean, the PP is an extension of the public health message that prevention is better than cure. This author mentioned that even though the PP "will allow action before convincing evidence is secured, it is not science averse," the fact that measures can be taken before any information arrives seems to be a lot safer for this author. From this standpoint, precautionary measures present the opportunity to analyze environmental and health research strategies and methodologies without taking any major risk.[56] Another possible benefit derived from precaution is that, according to this author, PP seems to offer a possible solution to escape from the impasse created by reductionism approaches. It opens up a way to do research and development without major risk. However, PP is based on the idea that all technologies and substances are dangerous until proven safe. This means that to err in prevention is better than to risk in development something harmful. Therefore, a substance that might cause damage to the environment should not be released

---

[55]   COMMISSION OF THE EUROPEAN COMMUNITIES. 2000. Communication from the Commission on the Precautionary Principle. Brussels. http://ec.europa.eu/dgs/health_consumer/library/pub/pub07_en.pdf.

[56]   GRANDJEAN, Ph. 2004. Implications of the Precautionary Principle for Primary Prevention and Research. Annual Review of Public Health, 25:199-223.

until proven safe, which, by logical conclusion, means that oxygen should not be released into the environment until proven safe, something that science cannot do because it is unpractical.

The obligation of scientific proof is a requirement of the PP, but the onus of this proof has to have a departure point. This departure point is a question in our case, the question is: What has to be proved? Gradjean gives the example of Denmark and Sweden, countries that have a list of chemicals which are undesirable and accumulative in the environment, it must be demonstrated (evidence) that these chemicals are not bio-accumulative before they can be taken off the list.

HEALTH APPROACH

For the World Health Organization (WHO), the PP can be summarized as "to respond to health risks before significant harm has occurred"[57].

According to Van der Haegen,[58] the PP is a legitimate tool that can be used by decision-makers in circumstances in which they are faced with potentially-harmful effects on health, but there is scientific uncertainty concerning the nature or extent of the risk. David Appell[59] mentions that because science does not have the answer to global warming, loss of biodiversity and toxins in the environment, precautionary measures should be taken. Uncertainty with science is a fact; therefore under this panorama caution comes first and science second.

---

[57]   WORLD   HEALTH   ORGANIZATION. 2003. Application of the Precautionary Principle to EMF. http://www.who.int/peh-emf/meetings/Lux_PP_Feb2003/en/

[58]   VAN der HAEGEN T. 2003. EU view of precautionary principle in food safety. speech at American Branch of the International Law Association. New York, October 23-25. http://www.eurunion.org/News/speeches/2003/031023tvdh.htm

[59]   APPELL D. 2001. The new uncertainty principle. Scientific American, January.

RISK TOLERANCE LEVEL

The right of countries to set their own appropriate or acceptable level of protection should be consonant with the chosen level of protection for any specific activity. In this regard the level of protection should be established in advance and should be recognized that there are some new or emerging risks. On the other hand, the evolution of scientific knowledge may influence societal tolerances and its chosen level of protection. In this regard the scientific knowledge that should influence society decisions and knowledge is related to uncertainty. Searching for a consensus on the real meaning of uncertainty Murphy mentioned that "... uncertainty is the raison *d'etre* of science."[60] While societal values are keys in determining a chosen level of protection against risk, in all cases, sound scientific evidence is a fundamental prerequisite to applying the precautionary approach.

Also situations where there is no threat of serious or irreversible harm to human health safety, the environment or resource conservation should not be considered related to the precautionary approach.

Risk management carries within itself the invocation of the precautionary measures; risk management is ultimately guided by judgment and based on values and priorities. Canada implements the Precautionary Approach in science-based health and safety programs and environmental and natural resource conservation, both domestically and internationally. The application of the Precautionary Approach to science-based risk decision making is often driven by specific circumstances and factors. According to the Canadian government the Precautionary Approach is a legitimate and distinctive decision-making tool within risk management.

---

[60]   MURPHY D.D., NOON B.R. 1991 Coping with Uncertainty in Wildlife Biology. Journal of Wildlife. manage. 55(4):773-782.

In the risk management framework process for known risks, developed by the US Presidential/Congressional Commission on Risk Assessment and Risk Management,[61] analysis of possible options, clarification of all stakeholders' interests, as well as openness in the way decisions are reached, are prominent. This is a basis for the WHO which incorporates a precautionary vision of risk management for uncertain risks into the process. Thus, the WHO combines the attributes of risk management for known risks with the attributes of the management process for uncertain risks into a single, enhanced risk management process.

I.3.1. GENERAL PRINCIPLES OF APPLICATION

The precautionary approach recognizes that the absence of full scientific certainty should not be used as a reason for postponing decisions where there is a risk of serious or irreversible harm.

The scientific evidence required should be established relative to the chosen level of protection. Further, the responsibility for producing the information base (burden of proof) must be assigned. One must recognize that the scientific information base and responsibility for producing it may shift as knowledge evolves.

It should be recognized that it is impossible to prove a negative (e.g., to prove categorically that something will cause no harm, prove with absolute certainty that something bad might not happen or to prove that something is not harmful), but it is possible to demonstrate that "reasonable testing" was done with no evidence of harm.

---

[61] PRESIDENTIAL/CONGRESIONALCOMMISSION ON RISK ASSESSMENT AND RISK MANAGEMENT. 1997. Symposium on a Public Health Approach to Environmental Health Risk Management. http://www.riskworld.com/riskcommission/Default.html

1.3.1.A PROPORTIONALITY

While some agreements, such as Rio, take in consideration the cost/benefit analysis and the possibility of applying
the PP according to the capacities of each country, other
protocols, such as the Bio-safety Protocol,[62] do not seem to
impose cost/benefit analysis. Cameron, in this regard mentions that the Bio-safety protocol is a stronger formulation
of the PP as stated in the Rio declaration, since in Articles
10(6) and 11(8) does not seem to require a threat of serious
or irreversible damage, or impose a cost-benefit analysis in
order to apply it [63]. But if precautionary measures should
be proportional to the potential severity of the risk being
addressed and to society's chosen level of protection then it
seems to be strictly necessary to take in consideration the
potential risk associated to the measures to be taken as well
as cost/benefit analysis. It is wise therefore to take into consideration, as stated at Rio, the cost/benefit analysis since
without it the proportionality between the measures taken
and the eventual risks associated to any event could have not
proportionality.

In other words, there is an implicit obligation to identify,
when possible, both the level of society's tolerance for risks and
potential risk-mitigating measures. This information should
be the basis for deciding whether measures are proportional
to the severity of the risk being addressed and whether these
measures achieve the chosen level of protection, recognizing
that this level of protection may change.

---

[62]   CONVENTION ON BIOLOGICAL DIVERSITY.
The Cartagena Protocol on Biosafety. http://www.biodiv.org/
biosafety/protocol.shtml
[63]   CAMERON J. 2001. The Precautionary Principle in International Law, in Reinterpreting the Precautionary Principle. T. O'Riordan, J.
Cameron and A. Jordan (ed.). Cameron May, London. p. 113-142.

1.3.1.b. Justice

Precautionary measures should be non-discriminatory and consistent with measures taken in similar circumstances. Consistent approaches should be used for judging acceptable levels of risk. Ultimately, the chosen level of protection should be established in the public's interest by weighing potential (or perceived) costs and benefits of assuming the risk in a manner that is consistent overall with societal values.

Where more than one option reasonably meets the above characteristics, the least trade-restrictive measure should be applied. This is because regulatory actions almost always have an economic impact on activities and precautionary decisions will almost always have a selective impact on them. Less trade-restrictive considerations should apply to the consideration of both domestic and international agreements.[64]

1.3.1.c. Cost/benefit.

The utilitarian approach would be to reduce exposure until the cost of the last reduction equals its benefit. However, society may wish to err on the side of caution and incur greater costs, in excess of the expected benefit. This may be the case for all risks, but is particularly relevant as an insurance policy against a small risk of a serious consequence, or in circumstances involving involuntary exposure, exposures to children and to certain diseases. Any effort to buy an insurance policy through risk regulation should ensure that the policy does not, itself, increase or create new risks. It is assumed that the final assessment of the benefit-cost effectiveness analysis will be performed at a society level.

---

[64]   Idem.

I.3.2 UNCERTAINTY AND DECISION MAKERS.

The most common problem when dealing with regulatory agencies and industries is the unclear effects of human health and the environment.

An example that clearly illustrates this point is the tobacco industry, tobacco growers and cigarette factories, who, for many years, opposed the regulation of the activity because there was not enough evidence that smoking was causing diseases. Today, many diseases are known to be related to the habit of smoking. After several court trials and many years of debate on something that was fact, the tobacco industry is now being regulated not based in the PP, but based on the consequences of the lack of it.

A problem that "can not be tested for a solution is defined as a wicked problem"[65] for which there is no definitive formula for a solution. In the case of ecological issues, these solutions do not belong to the field of ecology or explicitly to the so-called ecologist, since ecological problems create ethical, philosophical and legal issues. Essentially, ecological problems cannot be solved by ecologists, since they go beyond their boundaries.

According to Ludwig[66], Funtowicz[67] et al. who calls wicked problems post normal, they are characterized by radical uncertainty. Roe[68] calls these problems "truly complex" or "complex all the way down."

---

[65]   HOMER-DIXON T. F. 1991. On the threshold. Environmental changes as causes of acute conflict. International. Secur, 16: 76–116.

[66]   LUDWING D. MANGEL M., HADDAD B. 2001. Ecology, Conservation and Public Policy. Annual Review of Ecology and Systematics, (32) 481-517.

[67]   FUNTOWICZ S. O. MARTINEZ-ALIER J., MUNDA G., RAVETZ J.R., 1999. Information tools for environmental policy under conditions of complexity. Environ. Iss. Ser.: Eur. Environ. Ag. 54 pp. http://www.eea.eu.int

[68]   ROE E. 1998. Taking Complexity Seriously: Policy Analysis, Triangulation and Sustainable Development. Kluwer Academic, Boston.138 pp.

But there must be some reasons why these problems are so complex? One reason is that their solution does not belong to one field of science, and not even to strictly to science, instead to many other fields of knowledge such as social behavior including politics, religion, law, economics, international relations, law, moral values and ethical principles.

From Oreske's[69] point of view, when some ecological situations are poorly understood, when there are high population fluctuations or environmental stability is poor, some biologists have to make decisions. These decisions are based on uncertainty, not on scientific facts. In this context, statistically, there is always a grade of uncertainty, which must be evaluated in the light of the cost/benefit context.

On the other hand, there is always a temptation to make recommendations on the basis of the best available data in situations that are of critical importance[70]. This often means adopting a single best value for a parameter and a best hypothesis about the structure of the system. Such an approach may be misleading because it ignores the range of consequences that are plausible, but not excluded on the basis of available data[71].

But some scholars believe that a particularly harmful practice is to apply "hypothesis testing methods to management situa-

[69]   ORESKES N. SHRADER-FRECHETTE K. BELITZ K. 1994. Verification, validation, and confirmation of numerical models in the earth sciences. Science, 263: 641–46.

[70]   RUCKELSHAUS M., HARTWAY C., KAREIVA P. 1999. Dispersal and Landscape Errors in Spatially Explicit Population Models: a Reply Conservation Biology 13 (5): 1223.

[71]   MANGEL et al. 1996. Principles for the conservation of wild living resources. Ecol. Appl. 6: 338–362.

tions"[72] an affirmation also supported by Mange,[73] who mentions that human intervention in ecosystems is not a matter of just applying a hypothesis to test some statistical analysis. One reason for Mangel's point of view is that using mathematical models in a spreadsheet is not enough to make policy decisions; an analytic framework is needed to help structure the process of making environmental decisions.

On the other hand, it seems that intuition is unhelpful because environmental law concerns are outside our ordinary daily experience. Instead, decisions can be made as part of an ongoing series; our knowledge of the system can change in response to these decisions, rather than making them based on human knowledge[74].

It is well-known that uncertainty in ecological knowledge causes the extinction of dozens of animal species and exposes many others to extinction, including the American eagle, from the overuse of DDT, a chlorinate pesticide widely used for pest control in the middle of the 20th Century.

But what happens with regulatory agencies such as the Environmental Protection Agency the USA? At the beginning, in the early 70's EPA[75] was not concerned about the potential extinction of any plant or animal species. The reason: lack of knowledge of the causes of any human activity, but the lack of knowledge is not an enough reason not to regulate human activities as well as lack of knowledge or uncertainty should

[72]   LUDWING D., MANGEL M., HADDAD B. 2001. Ecology, Conservation and Public Policy. Annual Review of Ecology and Systematics: (32) 481-517.

[73]   MANGEL M. 1993. Comparative analyses of the effects of high seas driftnets on the Northern Right Whale Dolphin Lissodelphus Borealis. Ecol. Appl. 3:221−229.

[74]   PARMA A., AMARASEKARE P., MANGEL M., MOORE J, MURDOCH W. W. 1998. What can adaptive management do for our fish, forests, food and biodiversity? Int. Biol. 1:1626.

[75]   EPA. History. http://www.epa.gov/history/timeline/index.htm

not be a reason to stop economic or scientific development. Scientific knowledge or the lack of it is not enough reason to decide what should be done. Some philosophical and mathematical reason might establish the basis to apply precautionary measures under uncertain circumstances.

Understanding ecology scientifically is probably the most important tool for applying precautionary measures. We cannot know how nature is being or will be affected if we do not fully understand eco-systems. Science and mathematical models do not have the answer to all philosophical questions regarding science and scientific methods. Science deals with facts, not with values; therefore, it seems that scientific knowledge will never be enough.

Respect for human dignity and the biosphere, economic, ethical, political, religious and social insight must be considered before assessing whether human actions are beneficial or harmful to health and the environment. Also it seems that many problems, such as the depletion of world forest resources, endangered and threatened species and global climate change are not merely ecological or scientific. They involve a host of traditional academic disciplines that cannot be separated from issues of values, equity and social justice.

I.3.2.a- Scientific certainty

The grade of certainty is related directly to the grade of uncertainty; if certainty is 50%, uncertainty is 50%. The problem is that it is difficult to appreciate the grade of each problem in biological sciences, especially when factors such as exposure to hazardous substances, biotechnologies or policy options might affect human health in varying degrees.

It is widely accepted that the "better safe than sorry" attitude is warranted. The problem is: what is better safe? If this question has an answer, there is nothing to be sorry about, but uncertainty surrounds the "What is better?" In order to answer this question, Martuzzi [76] mentioned that decision makers ought to find the equilibrium between (i) costs and benefits, in terms of not only health, but also profit and economic return, (ii) using the available, often thin or worse, contradictory, evidence: (iii) winners and losers; (iv) immediate returns and long-term, perhaps hidden costs.

This author agrees that PP has the limitation of certainty and should be seen as a tool, but not the only tool, for decision makers to protect health and the environment.

The PP needs to be evaluated as both a principle and a tool, since, in itself; it does not provide a complete set of tools to address environmental and health problems. On the other hand, the concept of environmental factors should be extended further to those factors mentioned above.

Tickner[77] differentiates clearly between two major points of views of the PP: on one hand, the case of uncertain risk used to take preventive action, which is widely accepted in Europe, but less accepted in political circles in the United States. On the other hand, the portrait is painted of PP as anti-scientific, anti-innovation, or a "risk-management principle that is implemented only after objective scientific inquiry takes place." He disagrees with both of them, arguing that precaution is implicit in many governmental environmental and occupational health policies, but not as stated in PP. The problem, he argues, deals with uncertainty and the need for a strong scientific record before action can be taken. When

---

[76]   MARTUZZI M., 2004. Children in Their Environments: Vulnerable, Valuable & at Risk: The Need for Action. EEA/WHO/Collegium Ramazzini A one-day Science/Policy Workshop, Budapest.

[77]   TICKNER J.A. 2004. Children in Their Environments: Vulnerable, Valuable & at Risk: The Need for Action. EEA/WHO/Collegium Ramazzini A one-day Science/Policy Workshop, Budapest.

some scientific evidence is presented, those working on implementation of the PP move away from the notion of "invoking the precautionary principle" to show how science and policy can be modified so that decisions in the face of uncertainty are more preventive and protect health.

If the problem is improving decision-making under uncertainty, this author suggests four focuses to address the problem. There are: A) Alternative assessment —identifying, developing, and assessing a wide range of potentially safer alternatives to a particular activity or agent. B) Appropriate science —interdisciplinary approaches in scientific analysis that integrate qualitative and quantitative data and are more explicit about uncertainties. C) Increased participation and transparency in decision-making —expanding the range of constituencies, sources of knowledge and access to decision-making under uncertainty. D) Goal setting for environmental health— establishing broad health goals and implementing processes and metrics for reaching them.

In any case, the PP should be proactive, and should be in accordance with the preventive tradition of public heath. These groups of concepts should be consonant with the integral development of man and his environment, not the other way around. This means the main focus should be the development of humans, their health and, consequently, their niche.

Without doubt, the PP has been useful as a departure point to open up dialog between environmentalists and policy makers about scientific progress and human quality of life, but it should not be taken as an absolute principle or as a transitory tool to discover, in light of reasonable practical tools, concepts or principles to protect public health.

PP is particularly applicable to risk management, involving risk assessment and risk communication. Basically, the followers of the PP agree that if the risk cannot be determined with sufficient certainty, the PP should be applied. Another rule to regulate the application of PP by the European community (EC) is that the implementation of an approach, based on the precautionary principle, should start with a scientific evaluation as complete as possible, and, where possible, the identification of the degree of scientific uncertainly at each stage.

I.3.2.B.-ACCEPTABLE AND UNACCEPTABLE LEVELS OF RISKS

These levels of risk are not necessarily scientific mathematical models. The level of risk seems to be related to political responsibilities because an acceptable level of risk for society is political responsibility. Decision-makers faced with an unacceptable risk, scientific uncertainty and public concerns have the responsibility to find answers. Therefore, all these factors have to be taken into consideration. In some cases, the right answer may be to not act, or at least not to introduce a binding legal measure. A wide range of initiatives are available in the case of action, going from a legally-binding measure to a research project or recommendation.

The PP is, thereby, a tool for avoiding possible future harm associated with suspected, but not conclusive, environmental risks. Under the PP, the burden of proof is shifted from demonstrating the presence of risk to demonstrating the absence of risk. It is the task of the producer of a technology to demonstrate its safety rather than the responsibility of public authorities to show harm. The purpose of risk assessment is to provide accurate and useful risk characterizations to risk managers, who then can decide what should or will be done to reduce or manage those risks. Risk assessment involves four steps: hazard identification, dose-response evaluation, exposure assessment and risk characterization. Hazard relates to a particular item causing a documented effect. Dose-response evaluation involves determining the relationship between the magnitude of exposure and the probability of the adverse effect.[78] It is tempting to think that scientists, acting purely as scientists, can make the risk determinations that would trigger the taking of precautions. If this were true then, perhaps decision-makers could remove value-laden politics from the fact-finding processes that ground the domestic regulation of health and the environment.

Risk characterization can be viewed as a "quantitative measurement of the probability of adverse effects under defined conditions of exposure"[79]. However, the value of scientific risk analysis has been widely studied by Walker. This author questioned science as a neutral arbiter when dealing with PP,[80] in his terms "science cannot be a neutral arbiter for triggering precautionary measures, because both making and warranting findings of risk require non-scientific decisions. Making a risk finding requires decisions about the meaning of risk of harm, about the meaning of any modifiers for that predicate, and about the degree of confidence asserted for the finding as a whole" the main reason being that risk of harm is not a scientific decision. Another major concern is that the Cartagena Protocol Risk Assessment[81] has several weaknesses, including the subjective, case-by-case analysis and subjective level of risk . It states that "Risk assessment should be carried out on a case-by-case basis. The required information may vary in nature and level of detail from case to case…" This means that politi-

---

[78]   Idem

[79]   NATIONAL ACADEMY OF SCIENCES. 2000. Genetically Modified Pest-protected Plants: Science and Regulation. National Academy of Sciences Press, Washington DC. 292 pp.

[80]   WALKER V. R., 2003. The myth of science as a "neutral arbiter" for triggering precautions.http://www.bc.edu/schools/law/lawreviews/metalements/journals/bciclr/26_2/04_TXT.htm

[81]   CARTAGENA PROTOCOL ON BIOSAFETY. 2000.Anex III.

cal intervention and bureaucracy can make risk assessment virtually unpractical because of the unpractical application.

I.4.The principle of cooperation

According to the Catholic Church the Principle of Cooperation is intrinsic in Christianity[82]. It calls for brotherhood between all men as stated by the Catholic Church "Christians should cooperate willingly and wholeheartedly in establishing an international order that includes a genuine respect for all freedoms and amicable brotherhood between all. This is all the more pressing since the greater part of the world is still suffering from so much poverty that it is as if Christ Himself were crying out in the poor to beg the charity of the disciples. Do not let men, then, be scandalized because some countries with a majority of citizens who are counted as Christians have an abundance of wealth, whereas others are deprived of the necessities of life and are tormented with hunger, disease, and every kind of misery. The spirit of poverty and charity are the glory and witness of the Church of Christ giving the Principle of Cooperation a special place in the relationship between the Creator and manhood. It calls to Christianity to establish an international order, for sharing the goods among Christian's countries.

In the teaching of the Catholic Church cooperation with the Creator is an honor, given by God to man "in the end, it is God himself who offers to men and women the honor of cooperating with the full force of their intelligence in the work of creation."[83]

---

[82]   SECOND VATICAN ECUMENICAL COUNCIL. 1966. Pastoral Constitution Gaudium et Spes. Number 88.
[83]   PONTIFICAL COUNCIL FOR JUSTICE AND PEACE. 2005. Compendium of the social doctrine of the church. 460. Libreria Editrice Vaticana, Vatican.

To cooperate is a moral obligation of human brotherhood; Cooperation is different from charity. While some people claim that charity only helps the government to escape from its responsibilities; cooperation seems to be the official word for government charity, such is the case with the U.S. Agency for International Development, which cooperates in various programs in the developing world.

Cooperation is, therefore, the rational obligation of helping those in need. For instance, it is normal to have a segment of any given rich country living in poverty, but poverty is relative to the country's wealth. Poverty in New York is different than poverty in Calcutta, for example. Therefore, countries like Costa Rica, in Central America, with the highest standard of living in the region, have the moral obligation to cooperate with its neighbors in the areas where it has some advantages, such as agriculture and technology. The Rio Declaration is a declaration about cooperation among nations with an emphasis on cooperation between rich and poor countries and developed and developing countries. It is not about personal charity, but political and social cooperation. Principle 27 of the 1992 Rio Declaration[84] establishes that: "States and people shall cooperate …in the field of sustainable development."

This declaration mentions that partnership is a key to achieve cooperation, and this partnership should be achieved through the "creation of new levels of cooperation among States."

Global environmental and developmental systems reaffirm the Declaration of the United Nations Conference on the Human Environment, adopted at Stockholm on June 16 1972, and seek to build upon it, creating new levels of cooperation among States. The Catholic Church call for cooperation among nations to deal with global environmental issues

---

[84]   UNITED NATIONS. 1992. Conference on Environment and Development. Rio de Janeiro, Brazil. Principle 27.

and "modern ecological problems are of a planetary dimension and can be effectively resolved only through international cooperation capable of guaranteeing greater coordination in the use of the earth's resources." [85]

There are different forms of developmental cooperation, one being direct cooperation between one country and another, or bilateral cooperation. Cooperation can be also channeled via international organizations to many countries. In this case, it is called multilateral cooperation.

The application of the Principle of Cooperation seems to be regulated in the same act, which focuses on sustainable development. It calls for healthy and productive human life, while giving freedom to countries to exploit natural resources in a way that "does not harm" beyond each country's natural jurisdiction.

In medical practice, the Principle of Cooperation is widely accepted, and it is also widely accepted that universal knowledge, such as new medical techniques, medicines and treatments, should be shared by countries. Therefore, the Principle of Cooperation is accepted as a moral obligation in cases such as assisting poor countries to get rid of tuberculosis, river blindness or malaria. If health begins with good nutrition, shouldn't the Principle of Cooperation be applied to the transfer of agricultural technology? Shouldn't the Principle of Cooperation be applied to environmental issues? Shouldn't poor countries pay a bill for environmental cleanliness proportionate to that of wealthy countries? If so, what principle should we apply in cases of urgent technology transfer, such as GM crops?

It is not easy to find a short answer for this entire question, and many of these questions will have not short answers

---

[85] PONTIFICAL COUNCIL FOR JUSTICE AND PEACE. 2005. Compendium of the social doctrine of the church. 481. Libreria Editrice Vaticana, Vatican.

because of the complexity of the topic, from the Principle of Cooperation this question certainly deserves an answer based on social justice and brotherhood. Certainly, applying the Precautionary Concept without taking into consideration the reality of life in countries like Haiti or Ethiopia goes against the Principle of Cooperation and against the Universal Declaration of Human Rights states that "Everyone has the right to a standard of living adequate for the health and well-being of himself and of his family, including food, clothing, housing and medical care and necessary social services, and the right to security in the event of unemployment, sickness, disability, widowhood, old age or other lack of livelihood in circumstances beyond his control" [86] and the Universal Declaration of Hunger and Malnutrition (UDHM)[87] states that: "All countries, and primarily the highly industrialized countries, should promote the advancement of food production technology and should make all efforts to promote the transfer, adaptation and dissemination of appropriate food production technology for the benefit of the developing countries and, to that end, they should make all efforts to disseminate the results of their research work to Governments and scientific institutions of developing countries in order to enable them to promote a sustained agricultural development".

Furthermore, this conference calls for cooperation to preserve natural resources because it takes into account the importance of the environment on human nutrition and wellbeing; "to assure the proper conservation of natural resources being utilized, or which might be utilized, for food production, all countries must collaborate in order to facilitate the preserva-

---

[86] UNITED NATIONS GENERAL ASSAMBLY. 1948. Universal Declaration of Human Rights, Art. 25(1)
[87] UNITED NATIONS. 1974. Universal Declaration on the Eradication of Hunger and Malnutrition. World Food Conference, Rome, Art. 8-10.

tion of the environment, including the marine environment. All developed countries and others able to do so should collaborate technically and financially with the developing countries in their efforts to expand land and water resources for agricultural production and to assure a rapid increase in the availability, at fair costs, of agricultural inputs such as fertilizers and other chemicals, high-quality seeds, credit and technology. Co-operation among developing countries, in this connection, is also important."[88]

This declaration invokes the Principle of Cooperation since it, according to UDHM industrialized countries, promotes the advancement of food production technology and promotes sustained agricultural development. This declaration, as well as Rio, calls for cooperation among developing countries but, sadly, the UDHM declaration is often omitted by proponents of the ban and moratorium on GM crops.

Without cooperation, the PP is senseless in dealing with the environment, since this declaration is about the application of the PP according to their capacities and calls for cost-effective measures. For countries like Haiti, it's likely that all measures are not cost-effective because the country is so poor that it cannot afford any precautionary measure, not even the education of its own people. The Rio PP relates directly to human health, the environment and their reciprocal relation. In the case of the ozone layer, for instance, most measures will be not cost-effective for poor countries because developed countries' measures are more effective in producing most of the chemicals that are affecting the ozone layer. On the other hand, growing GM crops can be a cost-effective measure in China to prevent the degradation of the environment. In Europe,

---

[88]   UNITED NATIONS. 1974. Universal Declaration on the Eradication of Hunger and Malnutrition. World Food Conference, Rome, Art. 8-10.

where GMO might represent a threat of serious or irreversible damage to the environment, cooperation seems to be the basis for discussion of any risk associated to human health and the environment.

In summary, the Rio Declaration states that we are the center of concerns for sustainable development and we are entitled to a healthy and productive life in harmony with nature. Therefore, all countries and all people should cooperate in the essential task of eradicating poverty as an indispensable requirement for sustainable development. Sustainable development cannot be reached without a global partnership to conserve, protect and restore the health and integrity of the earth's ecosystem. Countries should cooperate in sustainable development, including exchange of scientific and technological knowledge to enhance development, adaptation, diffusion and transfer of new and innovative technologies. Countries should cooperate to promote a supportive and open international economic system that leads to economic growth and sustainable development in all countries to better address the problems of environmental degradation.

CHAPTER II
THE HUMAN NUTRITION PROBLEM

2.1. THE ETHICAL PRINCIPLES.

The Nufield Council on Bioethics (NCB)[89] based its analysis on agricultural biotechnology on three main ethical principles: general human welfare, the maintenance of human rights and justice. The delegates found broad differences in the ways these issues needed to be dissected, from purely technical aspects to more complex issues, such as whether moving genes between organisms is unnatural. At the light of the Catholic Teaching biotechnologies could help to fight the starvation and malnutrition problem however, emerging biological technologies have to deal with political, social, religious, ideological and economical issues before they can be beneficial in achieving their aims.[90]

2.2. TRADITIONAL AGRICULTURE

Traditional agriculture is the result of thousands of years of evolution. It is also based on technology and plant breeding with selection of desirable genes. Population growth has put

---

[89]   NUFFIELD COUNCIL OF BIOETHICS. 1999. Genetically Modified Crops: The Ethical and Social Issues. Nuffield Council of Bioethics. London. 164 pp. http://www.nuffield.org.
[90]   VOLANTE R. 2003. Intervention by the Holy See at the 32nd session of the conference of FAO. Rome, December 3.

more weight on the yield of crops; therefore, intensive agriculture has taken the place of extensive agriculture and agriculture of subsistence.

The development of technology at the end of the 18th century and during the 19th century, as well as advances in genetic breeding, were responsible for yield increase. Never before in the history of man was agricultural land as productive as in the past centuries. However, as yield increases, so does the use of in-farm agricultural chemicals and dependence on fossil energy. Traditional agronomists as well as plant pathologists recognized that high-input agro-systems are not sustainable and are being depleted, since cultivable land and natural resources are limited. In other words, highly agrochemical dependence systems have failed because they not only pollute the planet, but they are also limited by the two factors mentioned above.

Sustainable agricultural systems can be developed by achieving a more rational use of natural resources. This includes using genetics in processes, such as the transferring of genes from wild, resistant plants to plant pathogens or plant pests and also using vectors, such as viruses or plasmid genes, which can be transferred to develop new transgenic crops. Therefore, GE crops are probably the next step to help solving both pollution and limited natural resources.

2.3. SUSTAINABILITY OF AGRO SYSTEM

Conserving, protecting and restoring the health and integrity of the earth's ecosystems was one of the aims of the Rio meeting. The immediate question is how to protect it while solving the problem of feeding people. The answer does not appear too difficult, the main resource that humans have is probably their brain. New biotechnological advances should be used for human and environmental benefit.

GE food will probably make farming more sustainable, it might help to solve the problem of world production, result in better nutrition and, consequently, better health, but GE food is not in itself the complete answer for all problems of starvation, natural resource management specially water, political, religious, social and economical models and problems will not be solved by introducing GE crops.

Agriculture, therefore, needs a new revolution, and this time the new green revolution seems to be the GE revolution. Pesticide-free food, organic food and environmentally-friendly, sustainable agriculture are mentioned as agricultural systems that do not represent a treat to consumers, farmers or the environment. GE are proposed as an alternative to produce in a more sustainable way, sustainability implies a less dependant system of external inputs and greater use of removable resources. But not everyone agrees with GE Foods, on the contrary many countries especially in Europe are not that open to GE crops. Can GE crops be called environmental friendly and could they then be incorporated as sustainable systems?

The paradox is that organic agriculture results from conventional agriculture's reaction to pesticides and chemical fertilizers. One of its goals is to use natural renewable resources. For instance, *Bacillus thuringiensis* (Bt) is a bacteria that produces toxins against Lepidoptera. Instead of using the bacteria itself the genes that produce such toxins can be genetically incorporated into the plants killing Lepidoptera that eat this GE crops. But insects might be developing resistance to Bt, and companies like Monsanto that are producing "Superior Potatoes," which already come with Bt genes that codify for deadly insect toxins might be threatening organic farming rather than helping it because of eventual pest resistance.

2.4. GM Crops: The moral concerns

Traditional agricultural management practices as well as most agricultural systems are without risk, however, the public's attention to risk has been focused more on biotechnology than on its alternatives[91] possibly because of potential consequences of GM crops on the environment and human health. On the other hand GM crops have many practical and potential social and economical advantages.

For instance, most genetically engineered plants have added value, such as resistance to herbicide damage, higher yields, pest and plant pathogen resistance and potential use in vaccine production, resulting in better profitability for growers and reduce use of some agrochemicals, especially insecticides. The ethical debate is based on the safety of these crops for humans, animals and the ecosystems. These ethical concerns include the allergens produced by GM crops, cost-benefit analysis of transgenic crops vs. non-transgenic crops and environmental risk associated with both transgenic versus non-transgenic crops. While growing GM crops can help solve the problem of malnutrition worldwide and reduce the use of pesticides and consequent pollution of the environment and humankind, there is the potential risk that these genes can modify wild plants or negatively impact biodiversity.

On the hand, major issues of GM crops is the priority of caring for the ecosystem rather than the urgency for increasing yield. In this sense the moral issues are the major concern of the Catholic Church as entered in the misuse of biotechnology and the subordination of science over man, since researchers should "truly use their research and technical skill in the

---

[91]   SHELTON A.M., ZHAO J.Z., ROUSH R.T. 2002. Economic, ecological, food safety and social consequences of the development of BT transgenic plants. Annual Review of Entomology, 47: 845–81.

service of humanity".[92] The Catholic Church is also concerned in the development of applied biotechnology in agriculture, in part because of the potential imminent harm to consumers and the long-term effects on ecosystems, for the Catholic Church " the care of the ecosystem is a priority as well as the preservation for future generation."[93] Nature itself has the value of being part of the creation entrusted to man for his use and care, nature has the value of beauty and should be preserved for future generations as a gift that we have received from our ancestors.

Regarding the use of biotechnology in agriculture, the issue brings up an open discussion that include politicians, philosophers, law makers, growers, environmentalists and general public. It also brings hope to producers because the experience is showing that they are more profitable than traditional crops but also GM crops brings hostility, in some cases, because of the potential side effects on the ecological systems. The issue should be attended in a integrated way. Therefore the moral concern goes in two directions "their consequences for human health and their impact on the environment" as mentioned by the Compendium of the Social Doctrine of the Catholic Church.[94]

Human intervention on nature to modify some properties or characteristics such as plant genetic modifications in order to improve yield seems not to be opposed to the teaching of the Catholic Church, the care for nature is a moral responsibility. This moral responsibility includes the intervention of

---

[92]   PONTIFICAL COUNCIL FOR JUSTICE AND PEACE. 2005. Compendium of the social doctrine of the church. 458. Libreria Editrice Vaticana, Vatican.

[93]   Idem.

[94]   PONTIFICAL COUNCIL FOR JUSTICE AND PEACE. 2005.Compendium of the Social Doctrine of the Church. 472. Libreria Editrice Vaticana.

man in modifying genetically plants and animals, this mod-
ification does not constitute an illicit human act because if
they are for the benefit of filling the needs of man. However
those human actions that cause intentional irreversible damage to the environment or man ought to be rejected. For the
Catholic Church nature is a gift from the Creator but nature
is not sacred as is human life, humans have a supremacy over
all creatures, nature is instead "entrusted to the intelligence
and moral responsibility of men and women"[95]. Entrusted to
the intelligent means entrusted *pro bono* of both man and the
environment.

In the acceptability or not of biological and biogenetic
techniques it is important to notice that at the light of the
teaching of the Catholic Church precautionary measures
should be taken in order to avoid negative repercussions in
the long haul, the Catholic Church calls to "evaluate accu-
rately the real benefits as well as the possible consequences
in terms of risks."[96]

THE CONTROVERSY

It seems that the impact of GM crops on the environment
caught the attention of the media just after Losey *et al's* report.[97] In 1999 these authors mentioned that Monarch Butter-
fly (MB) was in danger of extinction because their larvae that
feed on pollen from GM Bt corn were dying.

It is well known that normally these larvae feed only on
the leaves of milkweed plants, which are commonly found

---

[95]   PONTIFICAL COUNCIL FOR JUSTICE AND PEACE.
2005. Compendium of the Social Doctrine of the Church. 473. Libreria
Editrice Vaticana.

[96]   Idem.

[97]   LOSEY J.E., RAYOR L.S., CARTER M.E. 1999. Transgenic
pollen harms monarch larvae. Nature, 399: 214.

in both natural habitats and cultivated fields. The problems arouse because the farmers program of insect control in corn, in the US, includes the use of transgenic corn with Bt Cry 1 toxins, this toxins are poisonous to some Lepidoptera larvae, a common pest of corn worldwide.

Losev et al.[98] made all their research in laboratory conditions not under field conditions. In their laboratory tests, Bt corn pollen was dusted onto milkweed leaves and then given the leaves to Monarch caterpillars to eat. Additionally, caterpillars were fed on leaves dusted with conventional corn pollen or leaves without pollen. The caterpillars eating leaves dusted with Bt pollen ate less and grew more slowly. Over four days, nearly half of the caterpillars on Bt-dusted leaves died while no caterpillars died in the other two groups. Five replications of the treatments provided enough data to indicate that the results were statistically significant. Losey and his co-workers reported their findings in May 1999 but supporters of GM crops were skeptics about their findings.

Two years later some other researchers published more conclusive field data in the Proceedings of the National Academy of Sciences[99],[100] showing that Bt 176 produces toxic substances to Monarch. But the fact that Bt176 is toxic to Monarch larvae does not means that under field conditions the Monarch is exposed to dangerous doses of Bt toxins. There are two facts to take into consideration the first one is that these components are found in the pollen and the second one is that this

---

[98]   Idem.

[99]   HELLMICH R.L. et al. 2001. Monarch larvae sensitivity to Bacillus thurningensis purified proteins and pollen. Proceedings of the National Academy of Sciences, 98:11925-11930.

[100]   ZANGERL, A.R. et al. 2001. Effects of exposure to event 176 Bacillus thuringiensis corn pollen on monarch and black swallowtail caterpillars under field conditions. Proceedings of the National Academy of Sciences, 98:11908-11912.

specific kind of corn comprises less than 2 percent of North America acreage. The reason why Bt 176 corn pollen is toxic is because the promoter used in Bt 176 is very effective in causing Bt protein. The conclusion was basically that under natural conditions the risk of exposure to this specific kind of pollen is very low.

The benefits

Pesticide use in agriculture is the major pollutant produced by humans. Some are widely used for soil-borne pathogen control and are harmful to the ozone layer. Therefore, GE crops aim to reduce the need for chemicals in agriculture with the benefit of an increase in biodiversity and less damage to the environment, additionally most GM crops are more profitable. To illustrate this point, according to USDA data from 1996 to 1998,[101] a net decrease in herbicide use in the United States was observed while a increase in yield was observed. This is only a consequence of new GM crops such as glyphosate (Paraquat, Monsanto)-resistant soybean. Shelton[102] mentioned, based on updated data, that transgenic crops are showing a positive economic benefit by reducing cost and the use of other insecticides hazardous to human health.

The use of Glyphosate-resistant soybeans also increased as more farmers began using it. Glyphosate has the benefit of remaining in the environment less than other alternatives for weed control. For instance, in soybeans, glyphosete remains roughly 47 days vs. other herbicides that remain between 60

---

[101]   UNITED STATES DEPARTMENGT OF AGRICULTURE. 2000. Agriculture Fact Book 2000. Washington, DC. 314 pp.

[102]   SHELTON A.M., ZHAO J.Z., ROUSH R.T. 2002. Economic, ecological, food safety and social consequences of the deployment of Bt transgenic plants. Annual Review of Entomology, 47:845-881.

to 90 days. It seems to also have some economical benefits for the grower as well as for the environment and, finally, the consumer. As a consequence of environmentally-friendly herbicides in soybeans and other crops, a 10% net decrease in chemical application was observed during the same period of time, according to this report. Still, some scientists believe that more data is needed in order to evaluate the benefits of GE crops on the environment and the risks and benefits to public health and the global economy[103], [104]

Another benefit of GE crops is that they might have more nutritional value or the potential to add more nutritional value, as is the case with vitamin A.[105] This was evident with GE rice or "golden rice," [106] which allows the body to synthesize the last step of the pathway from ß-carotene to vitamin A. There are millions of people at risk for vitamin A deficiency, and Golden Rice is one step forward. Many other deficiencies might some day be solved, such as folic acid to prevent cancer and malformation in prenatal babies. Vitamin A deficiency alone causes blindness in 250,000 to 500,000 children every year, and about 60% of them die within one year[107]

---

[103]   SCHOLTHOF K.BG. 2001. The chimerical world of agricultural biotechnology: food allergens, labeling, and communication. Phytopathology, 91:524-26.

[104]   TASYLOR S.L. 2002. Protein allergenicity assessment of foods produced through agricultural biotechnology. Annual Review of Pharmacology, 42:99-112.

[105]   GRUSAK M.A., DELLAPENNA D. 1999. Improving the nutrient composition of plants to enhance human nutrition and health. Annual Review of Plant Physiology, 50:133-61.

[106]   YE X., *et al*. 2000. Engineering the provitamin A (ß-carotene) biosynthetic pathway into (carotenoid-free) rice endosperm. Science, 287:303-305.

[107]   CENTER FOR DISEASE CONTROL AND PREVENTION. 1999. Global disease elimination and eradication as public health strategies. MMWR, 48:154-203.

Some GE crops were modified using a strategy called pathogen-derived resistance.[108] In short, a virus gene or portions of a virus genome is used to protect against a plant becoming infected by a virus in the field or greenhouse. This is the case with papaya and squash.[109] This technique for protecting plants from viruses that are vertically transmitted by insects is also useful in expressing pharmaceutical grade proteins and edible vaccines. [110] [111]

Mycotoxogenic fungi are carcinogens, and among the species of mycotoxins producing fungus, we find *Aspergillus* , *Fusarium,* and *Claviceps.* Aflatoxin, fumonisins, and ergot alkaloids are secondary metabolites of these species of common fungus that can be found in corn and many other grains.[112] Blue or green mold on fruits are infections caused by *Penicillum spp.* They are very common in other fruits, such as grapes, in which mycotoxins are produced. The infection occurs long before the plant, fruit or grain is harvested, but it is more evident as post-harvest diseases, mainly because of poor environmental storage conditions, including high moisture concentration, which also provide good conditions for fungal infections.[113] [114]

---

[108]   SANFORD J.C., JOHNSTON S.A. 1985. The concept of parasite-derived resistance-deriving resistance genes from the parasite's own genome. Journal of Theoretical Biology, 113:395-405.

[109]   BEACHY R.N. 1997. Mechanisms and applications of pathogen-derived resistance in transgenic plants. Current Opinion Biotechnology, 8:215-20.

[110]   CRAMER C.L., BOOTHE J.G., OISHI, K.K. 1999. Transgenic plants for therapeutic proteins: linking upstream and downstream strategies. Current Top. Microbiology and Immunology, 240:95-118.

[111]   RICHTER L., KIPP P.B. 1999. Transgenic plants as edible vaccines. Current Top. Microbiology and Immunology, 240: 159-76.

[112]   PHILLIPS T.D. 1999. Dietary clay in the chemoprevention of aflatoxin-induced disease. Toxicology Science, 52:118-26.

[113]   Idem.

[114]   PAYNE G.A., BROWN M.P. 1998. Genetics and physiology of aflatoxin biosynthesis. Annual Review of Phytopathology, 36: 329-62.

Mycotoxicoses in humans may cause gastrointestinal cancer and alfatoxin B1, associated with liver cancer, has shown a causal link between fumonisins and risk for esophageal cancer in South Africa. The Food and Agriculture Organization (FAO) estimates that 25% of the world's food crops are contaminated with mycotoxins.[115]

GE crops such as maize may provide a strategy to reduce *Fusarium sp.* and *Aspergillus* sp., reducing the risk of cancer in humans associated to mycotoxins. This strategy seems to be simple, but effective —already, 18% of the corn grown in the US is resistant to the European corn borer and the corn rootworm beetle because the Cry proteins expressed in Corn transform with *Bacillus thuringiensis* genes (Cry).

Leaf Superior is a transgenic potato clone genetically engineered by Monsanto whose leaves produce a copy of a toxin naturally produced by *Bacillus thuringiensis* (Bt) that is toxic to the potato pest Colorado Potato Beetle (CPB)[116].

EXPANSION AND BARRIERS

In 1998, genetically altered seeds, mostly cotton, soybeans, corn and cereals, totaled 18 million hectares in the U.S. mainland alone. Adding genes to them protected these crops from weeds, insect and plant pathogens.

The United States of America Food and Drug Administration (FDA) decided that there is no need for labeling GM crops, assuming that GE food does not represent a treat for Americans health. FDA considered GM crops equivalent to non GM crops and decided that no special label would be

---

[115]   SCHOLTHOF K. BG. 2003. One foot in the furrow: Linkages Between Agriculture. Plant Pathology and Public Health. Annual Review of Public Health, 24: 153-174.

[116]   POLLAN M. 1998. Playing God in the Garden. New York Times Magazine. October 25,1998.

needed for this kind of food.[117] However, not all people think the same way; some writers argue that transgenic foods are dangerous and represent a treat to the environment, especially in Europe, where GE foods must be labeled.[118] But the question is: why label something when it is innocuous?

One of the arguments against GE food frequently mentioned by opponents of GE crops is that GE seed production is owned by international corporations, and users must pay for the license to grow them. This means that producers have to pay for every single plant. They do not have permission to reproduce them because genes are intellectual property. Furthermore, in some cases, the same seed is registered as a pesticide at the EPA. However, this is a very weak argument since most growers are already purchasing seeds. Besides that, growers will not purchase GE seeds if it does not represent a benefit to them, which means that cost-benefit analysis does not balance the introduction of GE to their farming systems.

The argument that GE seeds can escape from farms to the wild is also a very poor argument, for two main reasons: First, GE plants are not necessarily more adaptable or competitive in the wild. On the contrary, they depend on man for competitiveness. The second reason is that most of them are not found anymore in the wild. As mentioned above, traditional agriculture is highly dependent on inputs, such as fertilizers, pesticides, machinery and fuel that make it an intensive practice, putting a lot of pressure on natural, non-renewable resources. These chemicals also pollute the environment, and, at the end, represent more of a problem

---

[117]   VOGT D., JACKSON B. A. 2000. Labeling of Genetically   Modified   Foods.   http://www.ncseonline.org/NLE/CR Sreports/Agriculture/ag-98.cfm

[118]   DONAHUE A. M. A WTO-Friendly Model: The European Union's Labeling. Requirements for Genetically Modified Foods. http://www.biotech-info.net/donahue.pdf

than a solution to food production. GE plants might not present a solution to the multiple problems of agriculture, but they are undoubtedly a great advance toward more sustainable systems.

There are still more arguments against GE crops. It is mentioned that transgenic food makers, such as molecular biologists, are breaking natural breeding laws by introducing genes that are not found in closed related species or, worse, are not found in plants such as the case with Bt/potatoes, in which a bacteria gene is incorporated into a plant genome.

Furthermore, some scientists argue that molecular biology is still in the dark, since there is no way to know exactly where a gene is going to end up once the plant is transformed. In order to transform a plant, a vector is needed, in this case, *Agrobacterum tumefaciens*, and the Bt gene is put inside the genome of *Agrobacterium*. Then the potato genome is modified. But many aspects of gene expression are unknown and depend on whether or not the gene will do what it is supposed to do. Bt modified genes, used mostly for Lepidoptera control, is a biological insecticide that has been used by organic and "pesticide free crops" growers, and is also widely used in sustainable systems such as organic agriculture. Basically, Bt is limited to foliar use.

In spite of the growing opposition to GM crops, the FAO report of 2004 mentioned that 68 million hectares of cropland worldwide were cultivated with GM or biotechnologically modified crops -- about 5 percent of the world's total crop area – The report says that that area is growing at a rate of 15 percent annually.[119] The adoption of GE plants in deve-

---

[119]   FOOD AND AGRICULTURE ORGANIZATION OF THE UNITED NATIONS.2000. The state of food and agriculture 2003-2004 agricultural biotechnology meeting the needs of the poor? Rome. http://www.fao.org/docrep/006/Y5160E/Y5160E00.HTM

loping countries went from 14% in 1997 to 24% in 2000, almost doubling in three years[120].

Finally, another argument is that GE crops are not natural. GE crops, according to their opponents, are not found in nature and, therefore, should not be incorporated into organic agricultural systems. This is another very sloppy argument, since conventional farming is the result of human *"euagrogenesy,"* or selection of good genes that give some benefit to humans, such as higher yields. This is not ethically wrong, considering human domination over nature, and represents an advantage to human self-adaptation. Essentially, GE crops are the result of human intelligence.

2.5. LIMITING FOOD PRODUCTION: GE AND MORAL RESPONSIBILITY

The Cartagena Protocol on Biosafety will likely become the next multilateral environmental agreement with the potential to set trade interests and environmental concerns in opposition. It might face conflict with the World Trade Organization (WTO). This is highly probable because the U.S., one of the largest producers of living modified organisms (LMO) that is part of the WTO, has not signed or ratified the protocol. The Cartagena Protocol on Biosafety, adopted in January 2000 as a supplementary agreement to the 1992 United Nations Convention on Biological Diversity (CBD),[121] could soon become one of the first binding multilateral international agreements dealing specifically and exclusively with some of the challen-

---

[120]   JAMES C. 1999. Global status of commercialized transgenic crops. ISAAA Briefs, No. 17, Ithaca, New York, 65 p.
[121]   UNITED NATIONS CONFERENCE ON ENVIRONMENT AND DEVELOPMENT. 1992. Convention on Biological Diversity. 31 I.L.M. 818.

ges created by "modern biotechnology."[122] According to the Biosafety Protocol: "Modern biotechnology" means the application of In vitro nucleic acid techniques, including recombinant deoxyribonucleic acid (DNA) and direct injection of nucleic acid into cells or organelles". A second meaning is the "fusion of cells beyond the taxonomic family, that overcome natural physiological reproductive or recombination barriers and that are not techniques used in traditional breeding and selection" To date, the Biosafety Protocol has been signed by 102 countries and ratified by 46. The Protocol requires 50 ratifications before it can enter into effect.[123]

Among the various responsibilities imposed by the Biosafety Protocol on each contracting party, a significant number are affirmative obligations to regulate the trans-border movement of living modified organisms (LMOs). The incorporation of trade-related obligations has led to an ongoing debate as to whether the Biosafety Protocol will conflict with existing international trade agreements, particularly with the World Trade Organization (WTO)[124] and its subsidiary agreements. It seems that the relation between the Protocol and WTO agreements further complicated the negotiations and delayed the adoption of the Biosafety Protocol. The unsatisfactory resolution of this question in the Protocol's text, or, more specifically, in its preamble, has not aided in putting an end to the debate. Many predict impending conflict, while some suggest the possibility that conflict is perhaps being overstated.

But, besides the legal battle between the WTO and the Biosafety protocol, the basic moral problem is the obliga-

---

[122]   CARTAGENA PROTOCOL ON BIOSAFETY TO THE CONVENTION ON BIOLOGICAL DIVERSITY. 2000. art. 3(i).
[123]   CARTAGENA PROTOCOL ON BIOSAFETY TO THE CONVENTION ON BIOLOGICAL DIVERSITY. 2000. art. 37(1).
[124]   GENERAL AGREEMENT ON TARIFFS AND TRADE. 1947. Statute A- 11.

tion to cooperate with poor countries, where people will be negatively affected by such international agreements that might become laws.

Fortunately, on the other hand, as with any other kind of restriction, finding ways around legislation or rules is part of human nature. Without compromise to the PP, researchers are finding ways to feed the poor. For example, in the South American Andean Region, genetically modified potato clones that provide resistance to the main Andean potato crop pests, nematodes, has been studied. The researchers found that even when gene flows occur in wild potatoes, there is no harm to non-target organisms[125]. The sterile male was transformed to provide a genetically-modified, nematode-resistant potato and to evaluate the benefits this provides until the possibility of stable introgression to wild relatives is determined.[126]

The Nuffield Council of Bioethics (NCB) suggests that introgression of genetic material into related species in crop biodiversity centers is an insufficient justification to ban the use of genetically modified crops in the developing world.[127] In this sense, there are different points of view about the potential benefits of GM technology in the developing world. Among these factors is the possibility of growing crops in inhospitable areas, which would help in alleviating food shortages. Furthermore, GM crops would require less use of developed world technology in the form of pesticides and herbicides, thus reducing costs for the developing world farmer. GM crops would be vital in feeding the world's rapidly growing population and providing edible, plant-based vaccines.

---

[125]   CAROLINA C. *et al.* 2004. Environmental biosafety and transgenic potato in a centre of diversity for this crop. Nature, 432: 222 – 225.

[126]   Idem.

[127]   Idem.

But opponents believe GM crops would increase costs for developing world farmers, as they would be forced to buy new seeds each season. GM crops would be designed to produce sterile seeds or no seeds at all, creating more dependence on the developed world, particularly because some GM crops require specific pesticides. Also, GM companies might not be interested in poor countries because the market is too small for their products. It is well-known that starvation in some countries is a political, rather than an environmental issue, so increasing the capacity to grow crops would not solve problems of under-nutrition. Furthermore, opponents suggest that producing GM crops in developed countries is cheaper than in third world countries and would increase poverty in those countries.

According to Shelton, Zhao and Ro,[128] ethics is an area of study that is often not explicitly stated, but influences the acceptability of biotechnology. In this sense, the NCB[129] reviews several of the ethical issues which are products of development and application of agricultural biotechnology in world agriculture and food security.

The NCB took place at a time when there were demands for banning GE foods and placing a moratorium on plantings, but the delegates did not believe there was enough evidence of actual or potential harm to justify a moratorium on GE crop research, field trials or limited release into the environment at that stage. Most importantly, the panel members urged the development of "a powerful public policy framework to guide and regulate the way GM technology is applied in the United

---

[128]   SHELTON A.M., ZHAO J-Z, RO, RT. 2002. Economic, Ecological, food safety and Social Consequences of the deployment of Bt Transgenic plants. Annual Review of Entomology, 47:845–81.

[129]   NUFFIELD COUNCIL OF BIOETHICS. 1999. Genetically Modified Crops: The Ethical and Social Issues. Nuffield Council of Bioethics, London., 164 pp. http://www.nuffield.org.

Kingdom." This Council aims to ensure that public concerns were addressed, the Council also urged for an independent biotechnology advisory committee and to take into consideration scientific and ethical issues together with the public values associated with GM crops. In the U.S., the debate on the ethics of agricultural biotechnology has been led by Thompson[130] and Comstock[131], the latter of whose earlier writings were decidedly against biotechnology, but who is now a cautious proponent of GE technology.

Therefore instead of being pessimistic amid the legal debate of the Biosafety Protocol scientists are trying to accumulate data regarding the safety of the GM crops, however scientific data is not enough without political disposition and a logical debate based on the principle of solidarity, the Principle of Cooperation, the principle of responsibility and social justice instead of extremist and pessimistic attitudes toward the potential risk of damage to biodiversity. The paradox is that in the name of science unjust illogical laws and agreements are being approved without taking into consideration scientific and social facts about human health and nutrition.

No one is morally obligated to follow an unjust law. There is a moral responsibility[132] in health issues, and nutrition. Therefore, NCB is on the right track. The fear of collateral effects does not outweigh the moral responsibility of feeding the poor. If we accept the precautionary concept as a legal principle and the cooperation principle and the principle of responsibility, as

---

[130]   THOMPSON PB. 2000. Bioethics issues in a bio-based economy. In The Biobased Economy of the Twenty-First Century: Agriculture Expanding into Health, Energy, Chemicals, and Materials. National Agriculture Biotechnology Council, Ithaca, NY. 196 pp.

[131]   COMSTOCK G. 2000. Vexing Nature? On the Ethical Case Against Biotechnology.Kluwer, Boston, 312 pp.

[132]   CADORE B. 1996. L'experice bioethique de la responsabilite. Arte Fides Namur, Montreal, 201 p.

ethical principles then the precautionary concept will outweighs legally any other principle but it does not means that this concept will defend human live. Furthermore, no moral principle can go against human natural right, assuming life and freedom are basic human rights as stated in the declaration of human rights.

In other instances the Catholic Church has been accused of stopping the development of science and technology, on this regards the Compendium of the Social Doctrine of the Church[133] mentions that "the results of science and technology are, in themselves, positive" even more, the teaching of the Catholic Church considers that "science and technology are a wonderful product of a God-given human creativity, since they have provided us with wonderful possibilities, and we all gratefully benefit from them.[134]" In the case of molecular biology and biotechnology applied to agricultural sciences the Catholic Church consider them as a gift from the intellect given by the Creator, with what is called "proper application", It seems to be evident that the scholars of the Catholic Church are aware of the potential and irreversible damage of biotechnology to the environment. This technology could help to solve the problem of malnutrition and starvation "they could be a priceless tool in solving many serious problems, in the first place those of hunger and disease, through the production of more advanced and vigorous strains of plants, and through the production of valuable medicines".[135]

---

[133]   PONTIFICAL COUNCIL FOR JUSTICE AND PEACE. 2005. Compendium of the social doctrine of the church. Libreria Editrice Vaticana, Vatican.

[134]   JOHN PAUL II. 1981. Meeting with scientists and representatives of the United Nations University, Hiroshima (25 February), 3: AAS 73, 422.

[135]   JOHN PAUL II. 1982. Address to the participants in a convention sponsored by the National Academy of Sciences, for the bicentenary of its foundation (21 September). L'Osservatore Romano, English edition, 4 October, p. 3.

CHAPTER III
THE ENVIRONMENT

3.1. EXTINCTION AND NATURAL HISTORY

It is believed that the universe is 15 billion years old and that planet Earth condensed about five billion years ago. Also, some physicists believe that life on this planet began 2.5 billion years ago and that it will become gas in the next five billion years before becoming a big black hole. In other words, the first living organism developed from chemical evolution to form small vacuoles that eventually evolved into a very primitive, bacteria-like organism, then into more complex prokaryotes and, eventually, eukaryotes. Eukaryotes still have some vestiges of this primate world, such as mitochondria. They evolved into very complex organisms and about 200,000 years ago into *Homos sapiens erectus* or modern man. Natural history tells us two mains things regarding extinction. The first one is that extinction of species is something natural for two reasons: evolution itself (competitiveness) and natural disasters. The second thing we can conclude from natural history is that all forms of life will eventually disappear. However, nature makes nothing purposeless in vain[136].

According to Petrini[137]the environmental ethics in the Catholic Church is Theo centric and is based in the prin-

---

[136]   ARISTOTLE. 1916. Politics. Dent, London, England, p. 16.
[137]   PETRINI C. 2002. Bioetica, Ambiente, Rischio. Logo Press, Roma. P390.

ciple of responsibility. In the Catholic Theology nature is a gift from God as well as human life, nature is created by God "God saw it was good" [138] and is entrusted to man, nature is the place where God wants man to live, it is man's environment. But in the history of creation there are responsibilities given by the Creator to man, they ought to care for their environment.[139] The garden (the environment) that the Creator gave man is in itself the gift from which man has to obtain his food and all his need thru his own work.[140] The second Vatican Council extended these responsibility toward all things created by man "especially with the help of science and technology, man has extended his mastery over nearly the whole of nature and continues to do so". [141]

During the 19th century and the first half of the 20th century, philosophy was mainly focused on nature. It was not until the second half of the 20th century that the relationship between environment and human beings became a topic in classrooms. This was probably due to the environmental crisis of that era, resulting in the widespread use of certain insecticides in agriculture, such as DDT and other chlorinates. In 1966, Rachel Carson[142] wrote the book "Silent Spring". This book was a collection of assays on aldrin, deldrin and DDT, all insecticides that caused a lot of damage to humans and wild animals, such as the American eagle, The history of fears on new technologies can be divided between before and after Carson. Fear is not a good friend, fears based on supposition and beliefs are even worse.

---

[138]   Gen 1:4,10,12,18,21,25.
[139]   Gen 2:26-30.
[140]   Gen 2:15, 3:17.
[141]   SECOND VATICAN ECUMENICAL COUNCIL. 1966. Pastoral Constitution Gaudium et Spes, 33: AAS 58, 1052.
[142]   CARSON, R. 1963. Silent Spring, Hamish Hamilton, London.

In 1954, Carson wrote: "The more clearly we can focus our attention on the wonders and realities of the universe about us, the less taste we shall have for destruction." Her four books concern the relation between humans and the environment. Human beings are the only ones that can do something to preserve nature. Before the 50's, inorganic pesticides, such as arsenicals, chlorinated hydrocarbons and organophosphates, were used for pest control, and Carson emphasized that they were causing neurotoxicity, leading to death in human and animals. She also mentioned that cancer hazards from polluted waters would increase in the future. Carson emphasized future ecological imbalance and the need for using biological control or less toxic chemicals.

Lynn White, in 1967, published the paper "The Historical Roots of our Environmental Crisis". He suggested that the abuse of nature by man is the result of the anthropocentric position of the superiority of human beings over other creatures. There was the belief that "overpopulation of humans" causes an imbalance and was not sustainable in the long haul.[143] This position was supported by Paul Ehrlich in his 1968 book "The Population Bomb." [144] He wrote "The battle to feed all of humanity is over. In the 1970s and 1980s hundreds of millions of people will starve to death in spite of any crash programs embarked upon now. At this late date nothing can prevent a substantial increase in the world death rate..."

Erhlich believed that the United States was overpopulated and called for immediate action. He proposed adding contraceptive materials to all food sold in the United States. Later, he rejected this idea because it was politically and scientifically unfeasible. But Ehrlich's predictions failed to become reality

---

[143] WHITE L. Jr. 1967. The historical roots of our environmental crisis. Science 155:1207.

[144] EHRLICH P. 1968. The Population Bomb. Ballantine Books, New York, NY. p.11.

because his linear model was wrong. Certainly, populations will grow in the coming decades along with food production.

Others writers, like Routley, believe that the Western view of nature is chauvinist or absolutely anthropocentric. From this standpoint, human acts are good or bad only in relation to humans, not in relation to other living beings.

For Routley, living organisms have intrinsic value that is independent of their usefulness to humans. Therefore, nature has value in itself, not in relation to humans. This position was also held recently by Peter Singer. Singer wondered if nature has value in itself or only in relation to humans. Would eliminating air and water pollution, as well as preserving wildlife, have value only because of their affect on human health, or do they have a value in themselves?[145] He answers this question based on the beauty of wildlife and its value as a source of knowledge for scientists and cures for human illnesses.

There are two major facts that can be opposed to early environmental writers, one is that man is the only one that can do something to stop nature's destruction and second is that the moral obligation toward nature should be based on the beauty of wildness. The assumption that man does something is based on the supposition that man is the one that is causing such damages, that nature does not have natural cycles. The fear that the planet was overpopulated when only it has half of today population still remains but perhaps human intelligence is greater than human fear and preventive action on natural disasters can be taking before they occur. Indisputably it will be a lot easier to correct human action that might affect nature than to take preventive actions on disasters caused by natural cycles. The answers to the question "Can man stop natural cycles from happening?" is not; "Can man build an

---

[145]   SINGER P. 2000. Writing on Ethical Life. HarperCollins Publishers Inc. New York, NY. p. 90.

arch to protect plants and animal from natural extinction?". nfortunately we have to accept that in spite of our political and social policies we can not control everything, what we can do is to know the difference in what can be and what can not be done. For those things that we can do something about, like pesticide pollution, preventive measures should be taken, since prudence is a cardinal virtue, and for those thing that we can do nothing, like solar cycles that affect weather change, precautionary measures should be taken.

3.2. Environment: PP and climate change

We ought to separate belief from fact, the extrapolation of in-lab research and field research since the belief that human have a nefarious influence on climate dynamics is becoming an article of faith. [146] It seems that "some facts" about how nature influences weather change is dispelling the belief that human's activities are responsible for weather change. Fortunately politicians, journalists and scientists are more cautious when dealing with affirmative statements related to the influence of human activity on climate changes. Since new data is revealing a much lesser influence of human activity on climate changes, the movements and people who still support these beliefs or "articles of faith of anthropocentric climate change" have found in the PP a shelter to hide from reality. As mentioned by Gerhard and Yannacone,[147] "their Precautionary Principle requires current action to mitigate speculative future impacts regardless of the present consequences intended or unintended."

To apply the PP to weather change without taking into account the natural history of climatic cycles might mean

---

[146]   LEE G., YANNACONE V.J.Jr. 2004. Invoking a Real Precautionary Principle.http://www.techcentralstation.com.
[147]   Idem.

condemning humanity to darkness. It is believed that human action will have some effect on a multi-millennial, long term temperature decline, but also on a natural, short-term rise. A proper application of the PP requires that people of the world be made aware that processes are at work that may raise sea levels, flood lowlands, and gradually shift climatic zones northward. Alternatively, the earth may already be overdue to slide back into glacial conditions. To hold out hope that reducing human energy use can alter either scenario is to condemn humans to suffer the effects of climate changes, whatever they might be.

The precautionary principle demands that public policy prepare to meet either or both of these unfortunate alternatives. Public policy must focus on mitigating the inevitable effects of the climate change that will certainly occur, rather than hoping we will be able to stop it.

Most probably climates will change. natural history tells us that there have been warmer and colder times over varying periods of time. This seems to be the natural rhythm of the planet, and we must adapt to these inevitable changes.

3.3. Biological diversity and ecosystems

The word "biodiversity" appeared to be used for the first time by Walter G. Rose in the National Forum on Biodiversity in 1986. Wilson and Peter published the first book on the topic entitled "Biodiversity."[148]

The term biodiversity refers to the earth's richness and varied array of living organisms, "whose species, the genetic diversity of individuals and the ecosystems that they inhabit constitute

---

[148]   PETRINI C. 2002. Bioetica, Ambiente, Rischio. Logos Press, Roma. p. 396.

what is known as biodiversity".[149] Therefore the term ecosystem is necessary for the definition of biodiversity, the most accepted definition for ecosystems could be the one described by the World Resource Institute (WRI): "ecosystems are communities of interacting organisms and the physical environment in which they live. Ecosystems are not just assemblages of species- they are combined systems of organic and inorganic matter and natural forces that interact and change. Ecosystems are intricately woven together by food chains and nutrient cycles; they are living sums greater than their parts. Their complexity and dynamism contribute to their productivity, but make them challenging to manage." [150]

Biological diversity has to do with the nature of ecosystems. The biological world is highly complex and diverse. This diversity depends on the relation of biotic and abiotic elements and the interaction among living organisms.[151] Biological diversity is greater near the equator, and it decreases the deeper you go into the soil and the higher you go into the atmosphere. Therefore, as you move closer to the earth's surface and closer to the equator, this diversity increases.

Humans are part of ecosystems and our influence on biodiversity and ecosystems is as old as the human species however it seems that after industrialization, this impact accelerated as the world's population increased.[152] The ecocentrism and biocentrism as proposed by some environmentalist are opposed to the teaching of the Catholic Church because "it is being proposed

---

[149]   WATSON. R. *et al*. 1995. World Evaluation of Biodiversity. Summary for Policymakers. UNDP., p. 9-10.

[150]   WORLD RESOURCES INSTITUTE. 2000. Guide to World Resources 2000 – 2001. People and Ecosystems: the Fraying Web of Life. Summary. World Resources Institute (WRI), Washington D.C., US., p.3.

[151]   DICKINSON, G., MURPHY K. 1998. Ecosystems. Routheledge: New York. P131.

[152]   Idem.

that the ontological and axiological difference between men and other living beings be eliminated, since the biosphere is considered a biotic unity of undifferentiated value. Thus man's superior responsibility can be eliminated in favor of an egalitarian consideration of the 'dignity' of all living beings".[153]

THE COSTA RICA CASE

Costa Rica biodiversity is protected by several laws and international agreements; however one of the weaknesses in protecting biodiversity is the knowledge of what is being protected, such as the number of species, therefore the first step to protect a country's biodiversity is to have an inventory of the different species and the spatial location of such species. In 1989 the National Biodiversity Institute (INBio) was created. It's task was to preserve biodiversity in Costa Rica is based on the "save", "know", and "use" strategy. INBio is part of a integrated biodiversity conservation strategy, a private initiative that is gathering, processing, producing and sharing biodiversity information.

INBio founders believe that despite all the political and subjective arguments that often emerge when conservation issues are discussed, the final decisions should always be based on scientifically sound information.

Information by itself seems to be not enough, some "prerequisites are needed in order to have some impact on biodiversity, and they are: scientifically sound, up-to-date, representative, available on various scales, sufficiently basic to be used as building blocks for other applications and readily accessible.[154]

---

153   JOHN PAUL II. 1997. Address to participants in a convention on "The Environment and Health", L'Osservatore Romano, English edition, 9 April. p. 2.

154   MATAE., GAMESR. 2004. Biodiversity Information and Policy-making in the light of Costa Rica Experience. IPTS Report 84. http://www.jrc.es/home/report/english/articles/vol84/ENV1E846.htm.

3.4. THE CASE OF AIR POLLUTION

Preventive measures should be taken, especially when dealing with environment. The PP concept could be useful and should be used in the right way. Yet, Raffensperger and others insist that the PP promotes science by encouraging the search for alternatives to real or perceived harmful practices and products. "I think we're in a unique time," said Raffensperger. "Events like the anthrax scare and SARS are showing us that we need to learn how to better connect the dots. Why are animal and plant species dying off? Why are rates of learning disabilities, asthma and some kinds of cancer rising? We can't necessarily wait for science in its splendor to prove connections after the fact."

There are thousands of studies that suggest a strong links between air pollution and health problems, especially for the elderly, children and those with respiratory and cardiac problems. A great number of studies, including some by the government of Canada, the Ontario Medical Association and the Toronto Public Health Department, show that air pollution can lead to premature death, increased hospital admissions, more emergency room visits and higher rates of absenteeism.[155]

For instance, ground-level ozone is a colorless and highly irritating gas that forms just above the earth's surface. It is called a "secondary" pollutant because it is produced when two primary pollutants react in sunlight and stagnant air. These two primary pollutants are nitrogen oxides (NOx) and volatile organic compounds (VOC). NOx and VOC come from natural sources as well as human activities. NOx are nitrogen-oxygen compounds that include the gases nitric oxide and nitrogen dioxide and are produced mostly by burning fossil

---

[155]   CANADA CLEAN AIR ACT. http://www.ec.gc.ca/cleanair-airpur/Home-WS8C3F7D55-1_En.htm

fuels. VOC are carbon-containing gases and vapors, such as gasoline fumes and solvents (but excluding carbon dioxide, carbon monoxide, methane and chlorofluorocarbons).

According to different information reports from U.S. and Canadian governments, human activities are responsible for recent increases in ground level ozone. About 95 per cent of nitrogen oxides from human activity come from the burning of coal, gas and oil in motor vehicles, homes, industries and power plants. VOC comes mainly from gasoline combustion and the evaporation of liquid fuels and solvents.

Ozone not only affects human health, but it can also damage vegetation and decrease the productivity of some crops. Additionally, it can injure flowers and shrubs and may contribute to forest decline in some parts of Canada. Ozone can also damage synthetic materials, cause cracks in rubber, accelerate fading of dyes and speed deterioration of some paints and coatings. It also damages cotton, acetate, nylon, polyester and other textiles.

Another problem is airborne particles, which include microscopic particles that remain suspended in the air for some time. Particles can be both primary pollutants and secondary pollutants sent directly into the atmosphere in the form of windblown dust, soil, sea salt spray, pollen and spores. Secondary particles are formed through chemical reactions involving nitrogen oxides, sulfur dioxide, VOCs and ammonia. Particles give smog most of its color and affect visibility. Depending on the type of particles, the air can appear yellowish-brown, or even white. Like ozone, particles are believed to have adverse effects on vegetation and various synthetic and natural surfaces.

Nitrogen Dioxide (NO_2) is a toxic, irritating gas that results from all combustion processes, and sulphur dioxide (SO_2) is a colorless gas that smells like burnt matches. Sulphur dioxide can be chemically transformed into acidic pollutants, such as sulfuric acid and sulfates, which are a major component of fine particles. The main sources of airborne SO_2 are coal-fired power generating stations and non-ferrous ore smelters. Sulfur dioxide is also the main cause of acid rain, which can damage crops, forests and whole ecosystems. Carbon Monoxide (CO) is a colorless, odorless and tasteless gas that comes primarily from automobile emissions. Ammonia is another pollutant in smog.

Most of these environmental pollutants are products of the planet's natural organic and inorganic activities, such as volcanoes, plants, animals and all forms of unicellular or pluricelular living organisms that produce the so called "pollutants."

These pollutants might affect, in a great or small, direct or indirect way, human quality of life. However, none of these single pollutant products of such activities can be compared to tobacco. "Tobacco is an obvious example," said Raffensperger. "We knew in the 1940s that tobacco was deadly. We knew about the deleterious effects of tar and nicotine. We had all of this information and yet nobody could prove in a court of law that tobacco caused lung cancer because science hadn't yet found the biological mechanism, the precise explanation of how smoking caused disease." The precautionary principle, Raffensperger said, provides a remedy. It urges action –or inaction– when logic suggests either is the prudent course.

Chapter IV
Conclusions

4.1. General conclusions

It seems that frequently supporters of the PP believe that the efforts to implement preventive actions are limited by scientific uncertainty that serves as an excuse to put off preventive politics and measures. Many times governmental agencies can not take action because the scientific data is inconclusive, this may result in delayed action to protect the environment.

As result of the literature reviewed, in this essay, one can observe that it is widely admitted that more research is necessary to determine the real cause of illnesses, and the relation between the biosphere and health, in order to identify opportunities for the risk prevention at its source.

It gives the impression that for some scholars the problem arises when precautionary actions are taken based upon our limited knowledge of ecosystems. Also it appears that if wrong, useless or ineffective actions are taken, these actions or omissions might harm human health.

Given that, in some instances, it seems that many actions have to be taken in light of poor knowledge. In the search of a friendly relation between health, PP and the environment I do agree with Martuzzi [156] when he mentions that the WHO is

---

[156]   MARTUZZI M. 2004. Children in Their Environments: Vulnerable, Valuable & at Risk: The Need for Action. EEA/WHO/Collegium Ramazzini , Workshop, Budapest, 22 June.

committed to "developing strategies and tools for supporting the development and adoption of policies in environment and health." In short these strategies, according to Martuzzi, must be practical and meaningful while mindful of all society's needs. The author admits that this strategy should be based on scientific evidence available on effects of environmental determinants on health. Strategies must also be equitable and sustainable.

It appears that the practical application of these WHO strategies might have remote health inference, partially because of the changes occurring in biotechnology, technology and political and social order.

Nevertheless the WHO seems to expand the scope of environmental and health topics, as it moves from the ground of environmental exposure and risk factors to the field of human health and environmental exposure reaching a new stage of "health determinants"[157].

One can conclude that the notion of environment is not limited to ecological issues; it goes into the fields of anthropology, philosophy, sociology and psychology. More specific it goes into "social, familiar and political context of children's lives"[158] furthermore , it goes into the contact with toxic agents. As a result, it could be accepted that the omission of the physical agents as the main factor that influences human health might mislead to wrong policies.

It is a fact that some particular human actions might cause environmental harm, and that environmental induced-imbalance might also cause harm to human health.

It seems that in seeking the fair approach to the environmental induced-imbalance on human health, leaving human health aside and focusing only on the biosphere ecological imbal-

---

[157]   Idem.
[158]   Idem.

ances issues is a major imprudence. First it creates a knowledge gap and second leads to "tremendous risks and confusion in the methodological path" says Martuzzi. From this perspective should the PP be used as a mere experimental tool when testing the management of health determinant? According to Martuzzi apparently yes, but some other opinions should be taken into consideration. Philippe Grandjean, *et al.*[159] takes us into the paradox of the PP. Their position seems to be reasonable, they support the fact that that "scientific uncertainty has been used as a pretext for postponing preventive efforts, while risk assessment debates have resulted in delayed action to protect public health and the environment". In his study, under the auspice of Collegium Ramazzini Grandjean, he concludes that: "A. In order to identify opportunities for the prevention of risks at the source, potential implications for health and the biosphere resulting from pollutants and other hazards, new research is needed. B. In order to take precautionary action to prevent ecological disaster, it is necessary to know the nature of the cumulative, complex, and synergistic effects on whole ecosystems. C. In light of the PP, scientific research and science policy need to be re-examined. D. In the spirit of PP, better links between science and public policy are needed. E. Scientific information should be used in decision making within a framework that facilitates the application of precaution to supplement prevention." [160] This author proposes that in some observational and experimental studies, based on the fact that statistical analysis results in more false negatives (error type II) than false positives (error type I) that tools and principles

---

[159]   GRANDJEAN Ph, *et al.* 2004. Implications of the Precautionary Principle in Research and Policy-Making (Commentary) American Journal of Industrial Medicine, 45:382–385.

[160]   GRANDJEAN Ph, *et al.* 2004. Implications of the Precautionary Principle in Research and Policy-Making (Commentary) American Journal of Industrial Medicine, 45:382–385.

of science, when applied to public policy, have often implicitly worked against precaution.

PP gives the impression of being an unsophisticated concept, seems like a little bit extremist principle, may be unpractical and perhaps it might change the way we get to know something scientifically, it goes into the philosophy of scientific knowledge.

Prevention of harm to human health can be addressed with specific methodology in order to evaluate the cost-benefit of intervention or not. Prevention might succeed but also prevention might fail at a highly environmental cost.

Medical ethics is based on four major principles: Beneficence, Non-maleficence, Justice and Autonomy. In this sense PP can be part of the principle of beneficence, also of no-maleficence, in this sense Weed Dean[161] mentioned that for WHO the PP is not considered a scientific, ethical, nor a philosophical principle but it could function as a legal principle. In his opinion the PP is only a guide: "PP, if it is a principle, guides how we apply scientific knowledge rather than how we acquire or conceptualize it." [162]For this scholar the action of prevention to avoid harm makes of PP an extension of the Principle of Beneficence.

Ethical choices ought to be evaluated in the light of reason; certainly a guide is needed and legal principles in order to implement those guides. Miller and Conko[163] calls the PP an "illusionary principle" from a "deeply flawed protocol," referring to

---

[161]   WEED DEAN D.L. 2000 Is the Precautionary Principle a Principle? In Special Issue on the Precautionary Principle. Refereed articles and Their Key Conclusions FOSTER K.B. VECCHIA P. (ed) IEEE Technology and Society, (21): 4.

[162]   Idem.

[163]   MILLER H., CONKO G. 2000.The Protocol's illusionary principle. Nature Biotechnology, 18: 360-360.

the PP as a "precautionary approach." [164] But the Precautionary Approach might be less controversial and more reasonable without loosing its significance. A Precautionary Approach is a lot better than PP, it might be more practical and have more ethical and philosophical significance. For Miller and Conko the PP might not be a principle. In fact, it does not even seem to be a clear concept and probably is not a total valid scientific tool to evaluate potential damage to the environment and, consequently, to human health. Furthermore, as a concept, it must be prudent to accept it with precaution. Sometimes national and international legislation rules morality, and usually behind most problems there are moral issues. Probably many problems arise when politicians try to convert moral choices into legislation. It makes sense to think that some politicians might have a need to legislate all fields of human life, and this need "leads to the temptation to find one absolute and easily applicable principle".[165]

It is easy to observe that there is more than one definition for PP also it is difficult to find a widely accepted definition; in some cases like the European Union there is not one definition for PP but several definitions. What seems to be true is that all interpretation of PP leads to precautionary measures. For some authors the PP is a "dangerous idea with the power over science and technology." [166]

We can find basically three positions on the PP: A) the extremist position that calls PP an "illusionary principle", B) the middle position that takes it with precaution and softens the possible negative effect of the PP on the economic and social development as well as the undesirable consequences for scientific and technological discoveries. For this middle position precautionary

[164]   Idem.

[165]   HOLM S, HARRIS J. 1999. Precautionary principle stifles discovery Nature, 400: 398.

[166]   MILLER H., CONKO G. 2000. The Protocol's illusionary principle. Nature Biotechnology, 18: 360-360.

measures should be taking according with the level of protection and taking into consideration the Principle of Proportionality.[167] And C) the position of the extremist environmentalist.

4.2. Cost-effective measures: Prevention of environmental degradation.

Environmental changes and degradation seem to have two sources, natural cycles and human activity, But before getting into some conclusions on the prevention of environmental degradation some questions arise; What protects the environment? Is protection widely applied by countries according to their capabilities? What are the cost/effective measures?

Natural environmental cycles can not be prevented, what can be prevented are some of the probable effects on human health, economical and social effects. On the other hand, the effect of human activities on the environment could be stabilized and make the systems sustainable. The spirit of the meaning of Rio is that environmental degradation has to be according to the capabilities of the countries leaves environmental protection to consciousness and what is cost-effective for one country might not be for it neighbors.

In some instances PP seems to be adopted as a dogma, without meditating on the practical consequences. But, Is the mere adoption of the PP the solution for the environmental degradation and the negative consequences for human health? It does not look like it, air pollution, the ozone layer, biodiversity issues and the endless number of human diseases associated with the environment will not be solved just by adopting

---

[167]   EUROPEAN COMMUNITY. 2000.Commission adopts Communication on Precautionary Principle. www.europa.eu.int/rapid/pressReleasesAction.do?reference=IP/00/96&format=HTML&aged=0&language=EN&guiLanguage=en

the PP and accepting it as an ethical principle. The problem is more complex and requires not only international legislation but political awareness and interdisciplinary work.

In practical ethics the term PP is still ambiguous[168] , PP has already entered the local and international law, but still it seems to be a more practical juridical principle for lawmakers than a practical ethical principle to solve many complex biological issues because prevention of environmental degradation is related to economical growth as well as environmental pollution, toxicity, research design and risk assessment.

Probably one of the major risks of PP is the belief that, if universally accepted, the PP will stop many possible harmful events from taking place. If we apply the PP and if this principle fails to reach its goals then the idea might be a dangerous one. Another risk associated to PP is that by adopting an universal principle, according to the local capabilities, you make the principle dysfunctional especially since it aims to protect the global environment. The third reason why PP might not succeed is the cost/effective measures, what is cost/effective for Costa Rica might not be cost/effective for neighboring Nicaragua so, while Costa Rica might take cost/effective measures to protect San Juan River for Nicaragua this might not be feasible.

The Kyoto Protocol is one example of a lack of cost/benefit evaluation. It demands the reduction of gas emissions in developed countries at the expense of lowering the standard of living and quality of life in the developed world with no major improvement in the quality of live of those in the developing world. The problem is that the Kyoto Protocol assumes that such actions will have a positive impact on the quality of life without taking into consideration at what cost. Before taking

---

[168]   GRANDJEAN Ph, *et al.* 2004. Implications of the Precautionary Principle in Research and Policy-Making (Commentary) American Journal of Industrial Medicine 45:382–385.

any action, this protocol should evaluate what action should be taken and at what cost. Who should take such action and who will benefit? And, to what extent will such actions solve the problem?[169] Kyoto's objectives cannot be reached without decreasing access to energy and restricting energy use. The result of this practice will lead to a consequent reduction in food production. In order to reduce energy use. Therefore it will be necessary to reduce birth rate in developing countries. If energy use is reduced at the expense of population reduction in third world countries, that means almost no one will benefit.

Can preventive, cost/effective and universal measures be taken without affecting the energy and food production? In order to answer this question one must know the end of such measures. If the aim is to protect environmental degradation and consequently overall health, in the long haul, then the cost pay in the short term should be assumed by the ones that can pay and should be directed to protect the poorest people of the world. Therefore wealthy countries should pay the bills of the poorest countries, and then the application of the Principle of Cooperation should compensate the impact of the proposed measures. In other words, good will and cooperation can make feasible universal and cost/effective measures without affecting energy and food production.

4.3. PP and the prevention of harm

There are two distinct philosophies concerning the assessment and regulation of potentially harmful substances: (*a*) risk assessment, favored in the United States, which tries to balance risk with public health and benefits; and (*b*) the PP,

---

[169] LEE G., YANNACONE V.J.Jr. 2004. Invoking a Real Precautionary Principle. http://www.techcentralstation.com/021704F.html.

used in some international treaties and increasingly in Europe, which provides more emphasis on avoiding potential risk and less emphasis on potential benefits. Depending on one's viewpoint, the PP can be seen as unscientific and having vague and arbitrary guidelines. From this point of view PP could negatively affect trade, social development, scientific innovation, technology and not have an effect on protecting the environment.

How much do we need to know in order to implement a new strategy and to prevent harm at the same time? Most medical advances have being done without knowing fully the consequences of their use, for instance the discovery and use of penicillin. In agricultural system many biotechnological advances were reached without knowing the environmental consequences however researchers were aware of the potential benefits, such was the case of the Bt plants, and therefore the amount of knowledge has to do with both the chosen level of protection and the probable benefits.

The analysis of cost-benefit should not be simplistic, for instance the essential debate about the use of biotechnology, including the use of Bt plants, should focus on comparing this technology to existing or developing technologies in at least the following areas: food safety and human health, environmental compatibility (including the effects on non-target organisms, and water supplies), benefits and risks to the producer and consumer, effects on food systems and issues of social justice (a complex series of often important but hard-to-quantify issues). Additionally, one should examine each on a crop-by-crop basis.

Some participants in the conference that issued the Wingspread Statement[170] argued that PP does not necessarily say no to innovation. They argued that the PP requires research-

---

[170]   Idem.

ers to evaluate "honestly all the evidence and uncertainties and that PP does not instruct us to balance evidence in a specific way."

In this line, Holm and Harris,[171] however, sustain that the "PP balances evidence in a specific way". Raffensperger *et al*[172] disagree with that statement. According to the latter author, PP is useful when decision makers suspect that a course of action may have "harmful effects but are uncertain about their cause and possible extent." It demands more, not less, science in decision making, trusting on multiple lines of evidence from diverse disciplines. But this argument is not convincing to everyone, the reason: to require more science probably is possible, furthermore it is convenient. On the other hand, based on previous experience to demonstrate that something will not cause irreversible damage, to the environment or human health, is almost, if not, impossible. To assume that something will not be harmful might be as risky as assuming the contrary, then the problem is that it ends up in a vicious cycle.

Scientific evidence can be obtained only with methodological analysis. It means that the analysis ought to be evaluated honestly. But to evaluate honestly is subjective and honest evidence is not valid scientific evidence. Furthermore, to evaluate "all the evidence and uncertainties" is materially and rationally impossible. We can evaluate only to the extent of our knowledge and capabilities.

It does appear easy to wrap up the PP in order to make a practical concept out of it. But probably the Precautionary concept (PC) could work as a good departure point. The first question regarding PP is: what it is for? According to Rio, it is

---

[171]   HOLM S.,HARRIS. J. 1999. Precautionary principle stifles discovery. Nature, 400: 398.

[172]   RAFFENSPERGER C. *et al*. 1999 And can mean saying 'yes' to innovation. Nature, 401: 207 – 208.

to prevent the degradation of the environment (prevention of harm), while the WHO takes the health approach: "to respond to health risks before significant harm has occurred". But the environment is man's natural habitat, and environmental degradation seems to be affected by human activities, both in developed and developing countries and by natural cycles.

Therefore, it is not the biosphere itself that is important, it is human health, but human health can only be achieved if the environment is healthy, and this includes all living organisms. Countries around the world have different standard for what level of public health is acceptable. Some countries invest more in environmental protection than others, for instance Brazil and Costa Rica, while other countries like Haiti and the Dominican Republic do not care as much partially due to cultural boundaries. In prevention some of the questions that needs more exploration are What can we prevent and what we can not prevent from happening? At what cost can we prevent them?, some authors also wonder "How much can we prevent?".[173] The cost of prevention now could be a small percentage of the cost of prevention in five to ten years from now. What it seems to be evident is that it is necessary to take adequate, realistic, practical and universal measures to preserve for ourselves and for future generation something that we inherit, our habitat, and that this measures should take into consideration the principles of justice, cooperation, responsibility and proportionality.

The PC is widely used in human health, but before focusing on PC it is necessary to know the general status of public health worldwide, what part of the pie is related to human habitat and what part is related to politics, physiology, cultural and educational boundaries.

---

[173] RAFFENSPERGER C., *et al*. 1990. And can mean saying 'yes' to innovation. Nature, 401: 207-208.

The challenge is to prevent harm before it occurs and to deal with problems that we already have, such as the hole in the ozone layer, depleted marine fish stocks and climate changes. The problem is that preventative actions have to be related to risk assessment, and a lot of data is needed before any action is taken since risk assessment is a scientific process that estimates the probability and severity of adverse events.[174]

In order to deal with uncertainty, Murphy[175] mentioned two factors as central to produce solid, defendable science: First, the rigorous application of scientific methods and, second, the development of clear operational definitions for terminology. Again we observed in this position an objective approach of the PP, not based in emotions or subjective unjust analysis, but in scientific evidence. Another problem observed by Murphy is the lack of clear definitions or terminology that result in ambiguous concepts.

Something worthy of mention is that many management decisions are made on the basis of incomplete information, in part probably because of the lack of adequate methodology, the lack of economical and human resources and most probably because of the need to make a political decision. In this sense for Murphy it seems that all management plans and conservation strategies have properties that can be stated as falsifiable hypotheses and can be subjected to testing with empirical information and predictions from ecological theory and population simulation models.

According to Bart *et al.* the PP is a policy guiding risk management, the aim of this policy is the reduction of risk by providing a guide to risk management based on the results of

174   HAIMES, Y.Y. 1998. Risk Modeling, Assessment, and Management. Wiley, New York.
175   MURPHY D.D., NOON B.R. 1991.Coping with uncertainty in wildlife biology. Journal of Wildlive Manage, 55(4):773-782.

risk assessments".[176] For this author PP dictates that the mere conception of a possible ecological hazard, even in the absence of scientific evidence, requires precautionary interim measures until research enables full risk management.[177] This is a reductionism and actually is the aim of the proponents of PP.

On the other hand Sternheimer[178] disagrees with Holm and Harris that real progress originates from the "refusal to take a path that would threaten one's own moral choices and values"[179]. Real progress can only be reached with ethical standards, progress based on slavery is not real progress, and if science contributes to progress then scientific discoveries should be ethical. In conclusion science should be considered as a daughter of ethics. It is the goodness of a human act that generates ideas that lead to discoveries which benefits others. In this sense researchers discoveries should be ethical discoveries not only in the way they carry it out but in the potential applications and side effects.

4.5. Conclusions on the GMO

GM crops might not be solve the problem of agrochemical pollution or the problems of starvation but it probably is the most powerful biotechnological tool to reduce land use, water use, labor and the input of fertilizers, pesticides and increase the quality and quantity of food. GMO's could help to make the agro system more sustainable, of course the pollution problem and malnutrition are far more complex and it would be irresponsible and illusionary to believe that GMO's are the

[176]   BART G.J., DICKE K., DIKIE M. 2003. Bt crop risk assessment in the Netherlands. Nature Biotechnology, 21: 973 – 974.
[177]   Idem.
[178]   STERNHEIMER J.1999. How ethical principles can aid research. Nature, 402: 576.
[179]   Idem.

solution for the lack of water or food production. In other words, biotechnology is just one more tool.

We ought to separate production, distribution and malnutrition problems. Some of the problems are social, economical and political problem rather than biotechnological problems. Ironically some countries in Latin America produce 3 times more than their internal needs and some countries have up to 24% of its population with malnutrition, totaling 52 million people suffering from malnutrition just in this part of the world.[180]

According to this review the presumption that GM crops might harm human health is not right because on the contrary a ban to GM crops might be more harmful that not taking the right actions at the right moments. So far, after years of producing GM crops, there is no great concern in the Americas or China, where most of the GM crops are produced.

Also several authors in this assay seem to agree that to apply the PP, as declared, might be ambiguous. Who is going to take the responsibility for the benefits to human health or the environment that can result by the prompt introduction of GM crops? PP does not tell us what to do in this case as precious time can be wasting in not developing hundreds of GM under the suspicion of environmental harm. In this sense the teaching of the Catholic Church might be right since technological advances could be "priceless tool in solving many serious problems, in the first place those of hunger and disease, through the production of more advanced and vigorous strains of plants, and through the production of valuable medicines".[181]

---

[180]   VARGAS A. 2006. 52 millones de personas sufren desnutrición en América Latina. La Nación 17 Octubre. http://www.nacion.com/ln_ee/2006/octubre/17/aldea863042.html

[181]   JOHN PAUL II. 1982. Address to the participants in a convention sponsored by the National Academy of Sciences, for the bicentenary of its foundation. L'Osservatore Romano, English edition, 4 October, p. 3.

Also, if some sources of water are already polluted with agrochemicals and the ozone layer has been damaged, in part, by biocides used in agriculture, why don't we look to potential benefits rather than potential damages to the environment?

In terms of GM crops one might wonder if the PP an unjust principle. PP does not tell us what to do if the potential benefits outweigh the potential damage. Nutrition and health cannot be separated because good nutrition decreases child mortality and increases life expectancy. Good health begins with nutrition, and GMOs can improve human health by improving human nutrition. On the other hand, GM crops improve the quality of life by reducing the need for water in agriculture and reducing water, soil and air pollution. Therefore the application of a principle, in the name of justice, that causes injustice is wrong.

For instance, a ban on GM crops will probably aggravate threats to biodiversity and conservation. It concerns human health and the environment, and that is why countries like Canada, the U.S. and the WHO are cautious about invoking the PP because of their wider perception of the problem. If applied with the concept of precaution and the Cartagena Biosafety Protocol, a ban on GM crops will be counterproductive. The poorest of the world could be the ones who will ultimately pay the cost of any ban on GM crops since they depend on the research and technological transfers from rich to poor countries. Besides that, they are the ones who need an improvement in food quantity and value. For the time being, they might keep polluting their already poor environment, paying more for agrochemical input and, consequently, more for food. A reasonable precaution could be to accelerate research, development and commercialization of GM food, not the other way around. If a large part of the world already lives below the poverty level an imminent solution of the problem is needed. Any ban on GM crops will only worsen the world's poverty. I agree with Goklany[182] that "the rewards of GM crops greatly outweigh their risk." It is a mistake to accept GM crops without precaution, but it is a bigger mistake to stop their use".

One good examples is the Cogem survey[183] where he addressed the potential ecological effects of insect-resistant plants on multitrophic environments, which are defined as systems in which organisms using food sources from various levels including primary producers (plants), primary consumers (herbivores) and secondary consumers (carnivores) interact. However, it seems that any pest control methodology will affect multitrophic interactions, mostly negatively, and that careful balancing of the pros and cons of insect-resistant transgenic crops versus classical plant breeding strategies may ultimately favor the former rather than the latter. Such risk-benefit analyses, however, can only be made after years of painstaking research if it is to conform to the precautionary approach, which dictates complete understanding of 'all' risks. That is why GMO will take a long time to be cultivated in Holland.[184]then if we take all precautionary measures, precious time could be wasted at the expense of the poorest of the world, falling into the paradox: "escaping goblins to be caught by wolves!". [185]

[182]   GOKLANY M. 2001.The Precautionary Principle: a Critical Appraisal of Environmental Risk Assessment .Cato Institute. Washington DC. p59.

[183]   BART G., KNOLS J., DICKE M. 2003 Multitrofe Interacties in Genetisch Gemodificeerde Gewassen een Enquête ter Identificatie van Belangrijke Aandachtsvelden voor Ecologisch Onderzoek. Cogem deskstudies serie: ecologie van genetisch gemodificeerde gewassen. http://www.cogem.net/pdfdb/2003-1.pdf.

[184]   BART G J., DICKE K. DIKIE M. 2003. Bt crop risk assessment in the Netherlands. Nature Biotechnology, 21: 973 – 974.

[185]   TOLKIN J.R. 1987. The hobbits. Houghton Mifflin Company,London p. 98.

Glossary of Acronyms

Bt Bacillus thuringiensis
CAMBIA Centre for the Application of Molecular Biology in International Agriculture
CBD Convention on Biological Diversity (Biodiversity Convention)
CC Catholic Church
CGIAR Consultative Group on International Agricultural Research
COPUS Committee on Public Understanding of Science
DES dietary energy supply
DNA deoxyribonucleic acid
EC European Commission
EPA Environmental Protection Agency
EPO European Patent Office
EU European Union
FAO Food and Agricultural Organization of the United Nations
GM genetically modified
GMO genetically modified organism
PA Precautionary Approach
PP Precautionary Principle
PC Precautionary Concept
UK United Kingdom
UNCED. United Nations Conference on Environment and Development

US United States of America
USDA United States Department of Agriculture
WRI World Resource Institute
WTO World Trade Organization.

Index

LibrosEnRed Publishing House

LibrosEnRed is the most complete digital publishing house in the Spanish language. Since June 2000 we have published and sold digital and printed-on-demand books.

Our mission is to help all authors publish their work and offer the readers fast and economic access to all types of books.

We publish novels, stories, poems, research theses, manuals, and monographs. We cover a wide range of contents. We offer the possibility to commercialize and promote new titles through the Internet to millions of potential readers.

Our royalties system allows authors to receive a profit 300% to 400% greater than they would obtain in the traditional circuit.

Enter www.librosenred.com to see our catalog, comprising of hundreds of classic titles and contemporary authors.

Lightning Source UK Ltd.
Milton Keynes UK
UKOW02f0849290316

271079UK00001B/98/P